THORNTON'S TEMPLE OF FLORA

Thornton's Temple of Flora

WITH PLATES

faithfully reproduced from the original engravings
and the work described by

Geoffrey Grigson

with Bibliographical Notes by

Handasyde Buchanan

Collins
14 ST JAMES'S PLACE LONDON
1972

John Brinkley scripsit

This edition of Thornton's *Temple of Flora* has been designed and produced for William Collins, 14 St James's Place, London, by George Rainbird Limited, Marble Arch House, 44 Edgware Road, London.

It is a smaller version of the edition designed and produced for William Collins by George Rainbird and Ruari McLean in 1951, and so marks the twenty-first anniversary of the publication of the first book produced by the Rainbird company. The plates and the text have been printed at Deventer in The Netherlands by Van Leer, who printed the 1951 edition, and the binding has been executed by Van Rijmenam of The Hague. The calligraphic title-page is reproduced from the design drawn by John Brinkley for the 1951 edition. The plates are reproduced from the copy of the original 1807 edition in the possession of Lieutenant-Colonel Philip Bradfer-Lawrence by his kind permission.

First published 1951

ISBN 0 00 211836 X

CONTENTS

LIST OF PLATES

In Colour

In Duotone

LIST OF PLATES *(Concluded)*

Note

Dr Thornton's own contents leaf, reproduced as the third of the black and white plates, is misleading in several particulars. Plate III in the original list of contents reads 'Sexual System of Linnaeus' but no such plate exists, while no mention is made of 'Flora dispensing Her Favours on the Earth' which is found in complete copies of the book ahead of the plates listed I and II. In the Bibliographical Notes of the present book therefore Flora is given as Plate I, Aesculapius etc. as Plate II, and Cupid as Plate III, while the other plates follow in Thornton's order, the alternative plates being given the letter A – thus XI The Queen, XIA The Queen Flower. The exact title given on Thornton's original contents sheet is not always to be found under the plate itself. 'Cupid Inspiring Plants with Love' has no title on the plate but instead four lines of poetry by Charlotte Lennox, while Group of Hyacinths, Group of Roses, Group of Tulips, Common Passion Flower, Hirsute Stapelia and Indian Canna become respectively Hyacinths, Roses, Tulips, Blue Passion Flower, Maggot-Bearing Stapelia and Indian Reed. Our contents page follows the titles on the original plates, with the exception of the Cupid plate.

THE TEMPLE OF FLORA,
OR GARDEN OF NATURE

NO OTHER BOOK of English flower illustrations has quite the celebrity of *The Temple of Flora*. Naturally so, for its appeal, whatever may have been the aim of Robert John Thornton, is to the emotions and not to the botanical scientist or even to the scientific or painstaking gardener. To place the vast folio on the table, to open it for the first time, to savour the rich plates, is to enter, as though it were still around us, the extraordinary theatre of romanticism. In these days the *Pontic* Rhododendron is a commonplace of roadside and shrubbery, gaudy forms of the Persian Cyclamen *(Cyclamen persicum)* are familiar in the glasshouse, and the tubers of the Dragon Arum *(Dracunculus vulgaris)*, so immensely portrayed against the dark mountains, can be bought for a few pence and left to throw up their purple spathes in a warm sunny border in the most ordinary garden. A hundred and sixty years ago, when Thornton conceived the huge work of which *The Temple of Flora, or The Garden of Nature* was to be the climax, the mood of men's mind was exotic. The romantic took many shapes. In Thornton, there is no doubt, the shape it took was mainly sensational and dramatic.

Let us consider first the man. His father, Bonnell Thornton, a doctor before him, was celebrated as a wit and a fop. He had married well and had inherited a fair sufficiency of £15,000. The young Boswell was pleased when Bonnell Thornton took notice of him and asked him to call – 'which I did, and found him a well-bred, agreeable man, lively and odd'. The father died in 1768, when Thornton (the exact year of his birth is unknown) was an infant. He was an ingenious boy, passionate about natural history, who kept at school a small botanic garden and an aviary containing every species of English hawk, most of them cunningly captured by Thornton himself. So much at least one may learn from an article in the *European Magazine* of July, 1803, which reads as though it had been written by the man himself to forward the sale and fame of his botanical undertakings. 'His grandmother often used to say, that she disliked young Thornton, as he was always catching of insects and butterflies in her garden, instead of minding his books. Hence all his weekly allowance went to maintain his garden and menagerie.' His mother sent him up to Cambridge in the hope that he would be ordained. The head of his college told her that he would undoubtedly shine in the pulpit. But he wished to be a doctor, took his Cambridge M.B., proceeded to Guy's, and studied abroad before setting up as a practitioner in London in 1797.

His mother died, his elder brother died, so Thornton inherited the family fortune, to which we owe *The Temple of Flora*. Having erected a tablet to his brother in the cloisters at Westminster Abbey, he took his heritage and began to think more seriously of an immense work in many volumes which in scope, illustration, paper and typography would surpass anything in any other European country. As a young doctor, energetic and well qualified, though untested, he was already making his way. One hero of his worship was Dr Thomas Beddoes, the strange father of the stranger poet Thomas Lovell Beddoes; another was Dr Erasmus Darwin, the first of the Darwins. Both were men of distinguished mind, and Thornton quickly followed Beddoes in his curious system of trying to cure one disease and another by making patients inhale different gases. Both were writers and scientific popularizers who used a way of writing which combined a glittering classical pomposity with sentimentality and exoticism. 'Unknown to sex', wrote Dr Darwin in one of his scientific poems, 'the pregnant oyster swells'. It was from such examples that Dr Thornton derived his own sugary pomposity of prose and his execrable taste in the species of poem written for *The Temple of Flora*. If his sprightly father had been alive, he would have found it hard not to make fun of his child's taste in literature and the models it was formed upon.

It is not easy to be sure of the right title of Thornton's immense work which included *The Temple of Flora*, for the simple reason that Thornton himself was not sure and wavered in his inclusive titles as the work itself so pregnantly swelled. Thornton was not altogether clear in his thoughts. At times he called it a *Philosophy of Botany* in which were to be included several subsidiary titles and volumes, though in the end the *Botanical Extracts; or The Philosophy of Botany* became three separate volumes in folio. What we know as *The Temple of Flora* is in fact the concluding portion of a work announced to the public in 1797 as a *New Illustration of the Sexual System of Carolus von Linnaeus*, published through the years in parts with green covers and pink labels, and complete by 1807 as a 'New Illustration of the Sexual System of Carolus von Linnaeus: comprehending an elucidation of the several parts of the fructification; a prize dissertation on the sexes of plants; a full explanation of the classes and orders, of the sexual system; and the Temple of Flora, or Garden of Nature, being picturesque, botanical, coloured plates, of select plants, illustrative of the same, with descriptions'.

Most of the *New Illustration* (and the *Botanical Extracts*) is orotund compilation without any enduring scientific importance; but it is as well to realize Thornton's grandiose conception of the whole. Botany and the biological sciences were in the ascendant. Linnaeus, the great classifier of plants and living creatures, was not long dead. J. E. Smith (whom Thornton succeeded as lecturer on medical botany at Guy's) had purchased Linnaeus's herbarium and library; and four years later in 1788 had helped to found the Linnean Society. Great botanical works with coloured plates were published or coming out in parts, at home and abroad, such as Curtis's *Flora Londinensis* (1777–1787) or Smith and Sowerby's *English Botany* (1790–1814); and strange flowers were coming into cultivation from the four quarters of the world. Moreover Erasmus Darwin's poem *The Botanic Garden*, in its two halves of *The Loves of the Plants* (1789) and *The Economy of Vegetation* (1799), soon established itself as one of the popular works of the time. A quotation (from James Thomson) on the title-page of the *New Illustration* shows what Thornton was about:

Shall Britons, in the field
Unconquer'd still, the better laurel lose? –
In finer arts and public works shall they
To Gallia yield?

Thornton insisted again and again that he was engaged upon 'a *national* undertaking'. Much followed from that, including the fact that the *New Illustration* was dedicated in his mighty style to the Queen:

From the unbounded protection, *so liberally bestowed by an august* KING, *and the best of* QUEENS, *all the useful and ornamental Sciences, with the pleasing Arts of Painting, and Engraving, have reached their pre-eminence; nor have the English Nation less reason now to be proud also of their superiority in Type and Paper. Whilst the honourable Exertions of a great Nation have been lately concentrated to embellish and illustrate the Fancy of Poets, or Sacred and Historic Truth; the Science of Botany, advanced as it is by Linnaeus, and subsequent authors, and by the glowing imaginations of modern Poets* [i.e. Dr Darwin], *who have improved on Ovidian Metamorphosis, seemed, likewise, to have a claim to enlist the fine Arts into her service.*

The coloured plates as well were specifically dedicated to Queen Charlotte, consort of King George III – 'Her Gracious Majesty, the Bright Example of Conjugal Fidelity and Maternal Tenderness, Patroness of Botany and of the Fine Arts' – just as the later *Botanical Extracts* were dedicated to the Prince Regent.

Thornton's aim, then, was the national glorification of botany and the alliance of botany and the pleasing Arts of Painting and Engraving, an aim principally concentrated in the plates of *The Temple of Flora*.

While the parts of the *New Illustration* were coming out, Thornton must have spent much of his fortune on assembling all the paintings and drawings in what was known as Dr Thornton's Linnaean Gallery, which he exhibited in 1804 at No. 49 New Bond Street and for which he prepared a full catalogue. Sold at a shilling, this 'Account of Dr Thornton's Exhibition of Botanical Paintings . . . . with the Poetic Compositions made on the different subjects, and explanatory Notes' went into at least four editions. The exhibition was made into something of a raree show with the aid of 'a bower, about which are disposed foreign, as well as curious English, birds and butterflies'.

In a footnote Thornton remarks that '*My Philosophy of Botany, or Botanical Extracts, including a new Illustration of the Sexual System of Linnaeus,* and my *Temple of Flora or Garden of Nature* . . . are now publishing, dedicated, by permission, to Her Most Gracious Majesty, at Twelve Shillings and Sixpence per number. The present Collection of Paintings was for the embellishment, or illustration, of these works'. Apparently twelve and six was soon found to be too low a price, since the advertisement at the end of the fourth edition of the catalogue announces that the twentieth number had just been published at twenty-five shillings. In the exhibition there were more paintings than were engraved finally for *The Temple of Flora,* and a study of the catalogue helps to explain the variation in copies of *The Temple,* the quarto edition, and the plates which were made up in portfolios for the third class of prizes in the Royal Botanic Lottery. Eleven botanists were portrayed – first Linnaeus, then Dr Darwin, and then, before the flower pictures began, Dr Thornton himself three times over, in portrait, miniature and bust. There is no denying that vanity – the other botanists came after the flowers – was one of Thornton's attributes, although he remarked in the catalogue that his own drawing of a group of roses (the original of one of the rarest plates in the *Temple of Flora*) was introduced 'with extreme diffidence' and was 'designed only to express the intentions of the Author of The Temple of Flora, or Garden of Nature, in directing his Flower Pictures'. A necessary precaution, since the catalogue also claimed brazenly that in his 'National Work' which was now publishing, 'all the most eminent British artists have been engaged'. It is clear that in showing his Linnaean Gallery and in planning the *New Illustration*, or at least that part of it which was to be *The Temple of Flora,* Thornton had in his mind Alderman Boydell's Shakespeare Gallery which opened first in 1789, and Fuseli's Milton Gallery. He might have taken warning from both enterprises, for they ended in failure, though Boydell unlike Thornton could fairly claim to have employed the most eminent of artists. But at the moment Thornton's hopes were still high, and his pocket – not as deep as Alderman Boydell's – was still not empty. His connections were well established. Great botanists of the day supported him. Erasmus Darwin had lived just long enough to inspect the early parts of the *New Illustration* and the first of the coloured plates. His approval went into the catalogue: 'I very much applaud and beg leave to recommend to the attention of the Public, the superb picturesque botanical Plates, now publishing by Dr Thornton, which I suppose to have *no equal'*. Other commendations were added by Daniel Rutherford, Professor of Botany at Edinburgh, and Thomas Martyn, the Professor at Cambridge and the first champion of Linnaeus in England, whose lectures Thornton had gone to when he was an undergraduate. Two things told against Thornton, however – the war with France, and a decided change in taste. His minor poets wrote poems for him in celebration of the doctor himself –

Thornton! while polish'd Darwin tells
The loves of Flora's gaudy train,
'Tis thine to guard from time's decay
The fading glories of her reign –

and in celebration of the chosen plants; but all of them – Miss Seward the Swan of Lichfield, Pye the worst of all Poets Laureate, Charlotte Smith, Cordelia Sheeles or Skeeles, Samuel Hull Wilcock, and Dr Shaw F.R.S. – used a manner of elegant personification and apostrophe which was losing its appeal; and in the compilation of his text Thornton's prose was no better or more convincing.

Subscription for the parts was probably disappointing, as it had been eventually for the parts of Boydell's Shakespeare. Thornton at any rate apologized to his subscribers when *The Temple of Flora* was at last complete: 'had the times been propitious' there would have been seventy coloured plates, in fact more than twice as many. But war had persisted in its interference. 'The once *moderately rich'*, the apology went on, 'very justly now complain they are exhausted through *taxes* laid on them to pay armed men to diffuse *rapine, fire, and murder,* over *civilized* Europe.'

Still, the national undertaking was complete, except for the *Botanical Extracts* and minor accompaniments such as *The Grammar of Botany,* which had been coming out in three-shilling parts since 1804, a copy of each part going, gratis, to those who subscribed to the *New Illustration.* The *New Illustration* had been sent to the Queen and the Prince Regent, and to the Emperor of Russia who returned a ring through the President of his Imperial Academy in acknowledgement of such 'splendid and elaborate Botanical Works'. Ambition was almost fulfilled – and Dr Thornton's inheritance almost drained away. Splendid plates, splendid calligraphic title-pages, splendid typography, splendid expanses of Whatman's wove, fees for artists, fees for engravers, all added to more than the doctor's fortune could endure without the fullest return. He looked again to the precedents set by Alderman Boydell, whose national undertaking for the advancement of the arts, and attempt to show the world 'that we can print as well in England . . . as they do at Parma, Paris or Madrid', had also been threatened by bankruptcy in spite of royal patronage and a dedication to the king. Boydell's way of saving his firm after the expense of the Shakespeare Gallery, and the publication, part by part through nineteen years, of *The Dramatic Works of Shakespeare* in massive folio, had been a lottery authorized by Parliament. The tickets had sold, the Gallery and its contents had been the first prize, and Boydell's firm recouped itself by some £66,000.

That was in 1805. Could the success be repeated in 1811 or 1812? It was win, or ruination. Could picturesque botany and Thornton's prose be as appealing as 167 pictures by Reynolds, Romney, Northcote, Hoppner, Fuseli, Peters, etc. and as Shakespeare's poetry splendidly set out? With the help of his powerful friends, including the Royal Librarian who wrote a letter hoping that he might recover some of the expense he had 'so liberally incurred', the doctor obtained his authority. In May 1811, 51 Geo. III Cap. 103 received the Royal Assent – 'An Act to Enable Doctor Robert John Thornton to dispose of his Collection of Paintings, Drawings and Engravings, together with several Copies of Certain Books therein mentioned, by way of Chance'.

The wording of the Act gives much information: for instance that the pictures were being exhibited once more, this time at the European Museum in King Street, St James's, and that Thornton had planned his work, 'which might eventually prove a national Concern and Honour', as early as 1791, even before he took his degree at Cambridge; but that 'the Sources of Assistance' upon which he depended had 'entirely failed'. The 'Botanical Lottery, for the Promotion and Encouragement of Science' was to be drawn between August 1st, 1811, and December 1st, 1813. No more was to be raised than £42,000 by the sale of 20,000 two-guinea tickets. The trustees of the lottery (the most notable of the three was James Everard, Lord Arundell, who planted an American Garden around Wardour Old Castle in Wiltshire) were to destroy the copper plates within a month of the draw, in the interest of the prize winners; and referees were also appointed to certify the final suitability of 'certain articles and things composing the Prizes' which were not yet completed. No doubt these included the very inferior quarto edition of *The Temple of Flora* (1812), or at any rate 600 sets of these quarto plates 'richly coloured' and 'valued £15 each' which were to be prizes of the fourth class, the three folio volumes of *Botanical Extracts; or The Philosophy of Botany,* 199 copies of which, together with 199 copies of *The Temple of Flora* and the *New Illustration,* formed the prizes of the second class, 'each set valued £80', the *Flora Britannica* (1812) of which 2000 copies each worth £5 were prizes of the fifth class, and the 7000 copies of the *Elements of Botany* (1812), worth £3 apiece and forming the sixth class. That leaves two classes of prize, the third and the first, or Capital, Prize. The third class was to be made up of 200 portfolios of the plates in *The Temple of Flora*. The Capital Prize, following the precedent of the Boydell lottery, was the Linnaean Gallery intact – 'being *all* the original paintings of Flowers, Allegorical Subjects and Heads of Botanists' then exhibiting at the European Museum, together with 'a superb copy' of each of the books which formed the other prizes. The ten thousand prizes in all were estimated to be worth £67,000.

What followed is not exactly known. Early in 1812 all the prizes must have been ready. Beneath a cut of Flora wreathed with flowers in her temple, a lottery prospectus was published calling upon the Nobility and Gentry to buy their tickets – 'Britons! join Hand and Heart in promoting the Arts and Sciences of your Country, by the Immediate Purchase of a Ticket'. Lots were to be drawn before June 4th, 1812, 'provided' – and here doubt shrieks aloud in denial of doubt – 'the encouragement from the Public has been such, that one Third of the Tickets are disposed of by that time; which, from the present generous patronage the Public has already exhibited, Dr Thornton cannot entertain one moment's hesitation about'. The lottery prospectus is reproduced on page 60.

It is certain that enough tickets had not been sold by June 12th, or even six months later. Thornton, in fact, did not give notice of the draw, in the *London Gazette* as he was enjoined to do, until March 15th, 1813. It was to take place on May 6th with the drawing of the State Lottery at Cooper's Hall in Basing Street. As the time drew near, the registered lottery agents advertised the tickets fulsomely in *The Times*, and, since the advertisements continued to the eve of the draw, the poorest sale may be inferred. Thornton concocted one device and another. The Emperor Alexander, who had acknowledged the gift of the *New Illustration* with a ring, was a universal hero after Napoleon's retreat from Moscow in 1812, so every purchaser of a ticket was to be 'presented (gratis) with an emblematical Print containing a strong likeness' of the Emperor, 'the illustrious Patron of Dr Thornton's Works', 'a striking likeness of the Brave Cossack!' 'All the PLATES are to be DESTROYED . . . immediately after the Drawing', screamed one advertiser on May 1st, so that the prizes 'will rise considerably in value'.

On May 5th, the day before the draw, the Linnaean Gallery was reopened ('free admission to all properly dressed persons'), to show again the grandeur of the first prize, viz: the 'choice Gallery of Pictures so much approved of by Mr West', the President of the Royal Academy; and on the same evening Thornton gave a botanical lecture in the Argyle Rooms. His grandiloquent style is discernible again in the advertisement. A child could guess why Thornton was lecturing on May 5th, but the lecture was for 'the Benefit of the Sufferers in Russia'; and the Doctor would 'expatiate on the great merits of Linnaeus, explain his System, and exhibit the choicest Flowers of Europe, Asia, Africa, and America, and discourse on them, and display a light obtained from Plants, equal to that of the Sun'.

On Thursday May 6th, the lots were drawn, and Thornton must have gone home knowing that his ruin was irretrievable. Twenty-five years later, the memoir of Thornton in the *Gentleman's Magazine* stated that 'the results were not sufficiently successful to restore his fortune, and he was ever after a beggared man'. Thornton's era of grandiosity had come to an end.

He continued in practice, not always successfully. 'I hear', a friend wrote to John Linnell the painter, 'you still consult that top-sawyer Thornton. He has been a thorn in my side; but I endeavour to forget his unsuccessful and expensive experiments on my poor son'.

From time to time, Thornton produced modest compilations, one of which was to bring him another portion of immortality. After the splendours and miseries of *The Temple of Flora,* he had still to be associated, not very happily, with William Blake. It appears to have been John Linnell who introduced Blake to his family doctor, with the idea that Blake should design wood engravings for a humble venture of Thornton's penurious days – the third edition of his Virgil for the use of schoolboys. Blake did the work. Thornton and his publishers could make nothing of Blake's remarkable designs and wished to have them all re-cut by a more conventional hand. The story is told by Gilchrist in his *Life of William Blake* – how poor Thornton was argued out of it by Sir Thomas Lawrence, James Ward, Linnell and others, and how the blocks were used, but with an apology by Thornton that 'they display less of art than of genius, and are much admired by some eminent painters'. Thornton was playing for a safety which he had not always found. Blake did not forget his man. When in 1827 Thornton published his last book, a *New Translation of the Lord's Prayer,* well intentioned to freshen the prayer from vain repetition, Blake annotated his copy, poked especial fun at Thornton's remark that God and his works would be better appreciated when better telescopes were made, and wrote his own parody of 'Dr Thornton's Tory translation, translated out of its disguise in the classical and Scotch languages' (Thornton was of Scottish descent) 'into the vulgar English'. He accused Thornton of making God into the Goddess Nature; and Blake was not fond of nature worshippers, whether they were Wordsworth or Robert Thornton.

In 1837, Thornton died at his home in London, leaving, apparently, not enough property to dispose of by will, or to make it worth while for his heirs to take out letters of administration. Was he the Robert Thornton who failed in his gambling speculations on the funds in 1814 and had to dispose of a library of books 'ornamented with extensive foppery'? Is he the Thornton whom Keats jokes about in a letter of 1818 as the folio size of physician out of practice? Possible and probable. But Keats would have found much to devour in the plates of *The Temple of Flora* of an extravagance compressible and dignifiable into fine poetry. We must ask why, out of the wreckage of a somewhat pompous and artificial life, these plates are still so highly valued and so fantastically entertaining or so simply enjoyable.

It is not unfair to say that the flower plates are the work of Thornton almost as much as the work of the men who first painted the pictures from which they were engraved so skilfully. Say the worst, prick poor Thornton's vanity, admit his impulsive greed for fame and success, acknowledge that his taste was not always of the angels, scarify him for always preferring the bad poet to the good –

With honey'd lips enamoured Woodbines *meet,*
Clasp with fond arms, and mix their kisses sweet –

accuse him of being a compiler and a popularizer and no scientist, damn him in good earnest and pity him for a failure. Still there remain these plates, by the splendour of which he – and his artists – have their posthumous security. Thornton, as he claimed, *directed* his flower pictures. He chose the plants, arranged the symbolism and the backgrounds: 'Each scenery is appropriated to the subject. Thus in the *night-blowing* CEREUS you have the moon playing on the dimpled water, and the turret-clock points XII, the hour at night when this flower is in full expanse. In the *large-flowering* MIMOSA, first discovered in the mountains of Jamaica, you have the humming birds of that country, and one of the aborigines struck with astonishment at the peculiarities of the plant. In the *Canada* LILY there is expressed the shade it delights in, with a sky whose clouds yet contain snow within their bosom . . . The TULIPS and HYACINTHS are placed in Holland, where these flowers are particularly cultivated, embellishing a level country. The ALOE erects, in contrast, its stately form among mountains, and the height and shape of the whole plant may be seen in the background. In the *maggot-bearing* STAPELIA you will find represented a green African snake, and a blow-fly in the act of depositing her eggs in the flower, with the maggots produced from this cause. The clouds are disturbed, and everything looks wild and sombre about the *dragon* ARUM, a plant equally posionous and foetid'. His artists, Peter Henderson, Philip Reinagle, Abraham Pether and Sydenham Teak Edwards, had to obey. We can be sure of that; and, in giving the orders and exacting obedience, he did provide them with their share of fame, for none of the team are otherwise much remembered in the history of art. Henderson and Reinagle did most of the work, and neither of them is among the 'most eminent British artists'. Apart from his competent botanical work we know almost nothing of Henderson. Reinagle, after painful years of failure in the Academy elections, at last become an R.A. He was adept at portraits, sporting pieces, landscape, and copying. He was always short of money (so was Abraham Pether) and in 1798 was appealing to the Academy for £150 'to save him and his family from ruin'. Allan Cunningham wrote not very kindly of this versatile little man, the son of a Hungarian musician who came to Scotland with the Young Pretender, that he 'may be numbered with those who increased the quantity of the pictures, without adding anything to the original lustre of his school.' Pether and Sydenham Edwards count for less in *The Temple of Flora,* contributing only a total of three and a half plates. Edwards, a Welshman, does have an honourable and enduring place, independently of Thornton, for his share of the great *Flora Londinensis,* for plates in the *Botanical Magazine* and his own *Botanical Register.* Pether, versatile again, was a nature-romantic who shared with Wright of Derby that fancy for moonlight which made him the man to paint in a moony background for the Night-blowing Cereus (Plate XIV) which Reinagle conceived so dramatically. When he died soon after *The Temple of Flora* was completed, he left his children abjectly poor. So Thornton's painting team was not as distinguished as he always made out. They were no doubt the men whose services he could afford without too much cost. But they were professionals, well able to delineate a plant set before their eyes, and to give each plant the dramatic setting which Thornton required.

Why, exactly, do we like the plates and treasure them? Is it because they are masterpieces of art making us think of the flowers of Dürer or Hans Weiditz or the drawings of Leonardo, or even because they are superlatively good in the lesser realm of draughtsmanship applied to plants? I think not, and it is no disparagement to say as much. No, they are large, excellently engraved and coloured, they are sumptuous, they stir the feelings, sometimes wildly, sometimes amusingly, and they delight our sense of period. The world they belong to is that of Monk Lewis and Mrs Radcliffe and Erasmus Darwin and the fatal Upas Tree, of the Gothic folly and the classical temple side by side in the landscape garden; or more exactly of the classical temple above the serpentine waters of the lake, with the grotto, the rockwork, the cascade, the plantation of American trees nearby to stand for the complementary and horrid graces of the wilderness. The parallel is close. Think of the famous landscape gardens of Stourhead in Wiltshire. On one side of the winding lake you have indeed the very Temple of Flora classically contrived of limestone; on the other side you have the grotto through which the waters of the infant Stour tumble into the lake under the rough ceiling of *pierre antédiluvienne,* though the guardian of the waters is a leaden nymph by Rysbrack. Or King's Bowood in the same county. There, too, is a temple above the lake, and below it, beyond the dam, a wilderness of rocks and niches and a cave and cascade concocted by a master-builder of grottoes, Josiah Lane. Now come back to *The Temple of Flora* and, as you turn over the plates one by one, see how many such elements are combined. Here is romanticism, neoclassic, oriental, and Gothic – romanticism of sentiment and romanticism of horror. Cupid inspires the plants with love; Reinagle – or Thornton through Reinagle – perches a jungle parakeet over the fat little boy's head (Plate III), Pether paints the Persian Cyclamen with an Oriental temple in the background (Plate V). With the exoticism of the Night-blowing Cereus, first brought from Jamaica c. 1700, you have midnight, the Gothic tower, the oak and ivy and the light of the moon upon the water (Plate XIV). Behind the Indian Reed (Plate XXIX) stands a Chinese pagoda, as it might be by Sir William Chambers in the gardens at Kew, though Thornton states that the temple which breaks the dark background of the White Lily (Plate XX) is 'the only kind of architecture that can be admitted in the garden'. So, too, the three Passion-Flowers rise against fluted columns (Plates XVII, XVIII, XIX). Horror-romanticism is in the Maggot-Bearing Stapelia (Plate XXIII) and the Dragon Arum (Plate XXII). In the more sensational but inferior plates of the quarto edition, the Dragon Arum gathered lightning to its mountain back-cloth. In his text Thornton gave it a prose poem. 'This extremely foetid poisonous plant' he wrote, 'will not admit sober description. Let us therefore personify it. She came peeping from her purple crest with mischief fraught: from her green covert projects a horrid spear of darkest jet, which she brandishes aloft: issuing from her nostrils flies a noisome vapour infecting the ambient air: her hundred arms are interspersed with white, as in the garments of the inquisition; and on her swollen trunk are observed the speckles of a mighty dragon: her sex is strangely intermingled with the opposite! confusion dire! – all framed in horror . . . ' He remembers, though, his medical botany, and adds a note that from the root 'a powerful and useful sternutatory may be made'.

Most of the species illustrated had been familiar for many years. Others were introduced in the seventeen-seventies, -eighties, or -nineties, giving novelty, *quantum suff.*, but not too much, to *The Temple of Flora*. One plant *Strelitzia Reginae* (Plate XI), The Queen, or the Bird-of-Paradise Flower, which came from South Africa in 1773, was included as a compliment to Queen Charlotte (to whom the *New Illustration* and *The Temple of Flora* were dedicated), for it was named after her, as Charlotte of Mecklenburg-Strelitz.

To arrive at Thornton, boil down all the diversities of the time in architecture, poetry, painting, fiction, music, from Walpole and Sir William Chambers to Shelley and the music of Weber, by way of Chatterton writing in Gothic or African mood of medieval Bristol or reeking tigers, by way of the wild imaginings of Turner or Martin or Francis Danby, by way of elements even in Blake, in Coleridge, in Wordsworth. An age is packed into these folio plates which Thornton directed. To some of the diversity they look back, to some they look forward. Ahead, not to far ahead, quivers the Sensitive Plant of Shelley's poem, which 'opened its fan-like leaves to the light' as Reinagle painted it (Plate XVI), and which grew in the poem among many other flowers of *The Temple of Flora*. Further ahead of the three plates by Reinagle and Henderson (Plates XVII, XVIII, XIX), a splendid tear, in Tennyson's *Maud,* falls *From the passion-flower at the gate*.

If we add up all we know of Thornton and consider his career, his vanity and occasional absurdities, the pomposity and excess sugar of his writing, Keats ridiculing him, and Blake doing the same, we shall be forced to admit that he was, in some ways, a figure of fun. But, as he once quoted from Cowper, 'The spleen is seldom felt where FLORA reigns' and, to a figure of fun who left us *The Temple of Flora* and beggared himself in the making of it, everything else may be forgiven.

Turn quickly, then, from the weaknesses of the man to the carnations of Plate VIII, the tulips of Plate X, the white lily of Plate XX, or the pink Kalmia of Plate XXVII against its background of pale blue.

GEOFFREY GRIGSON

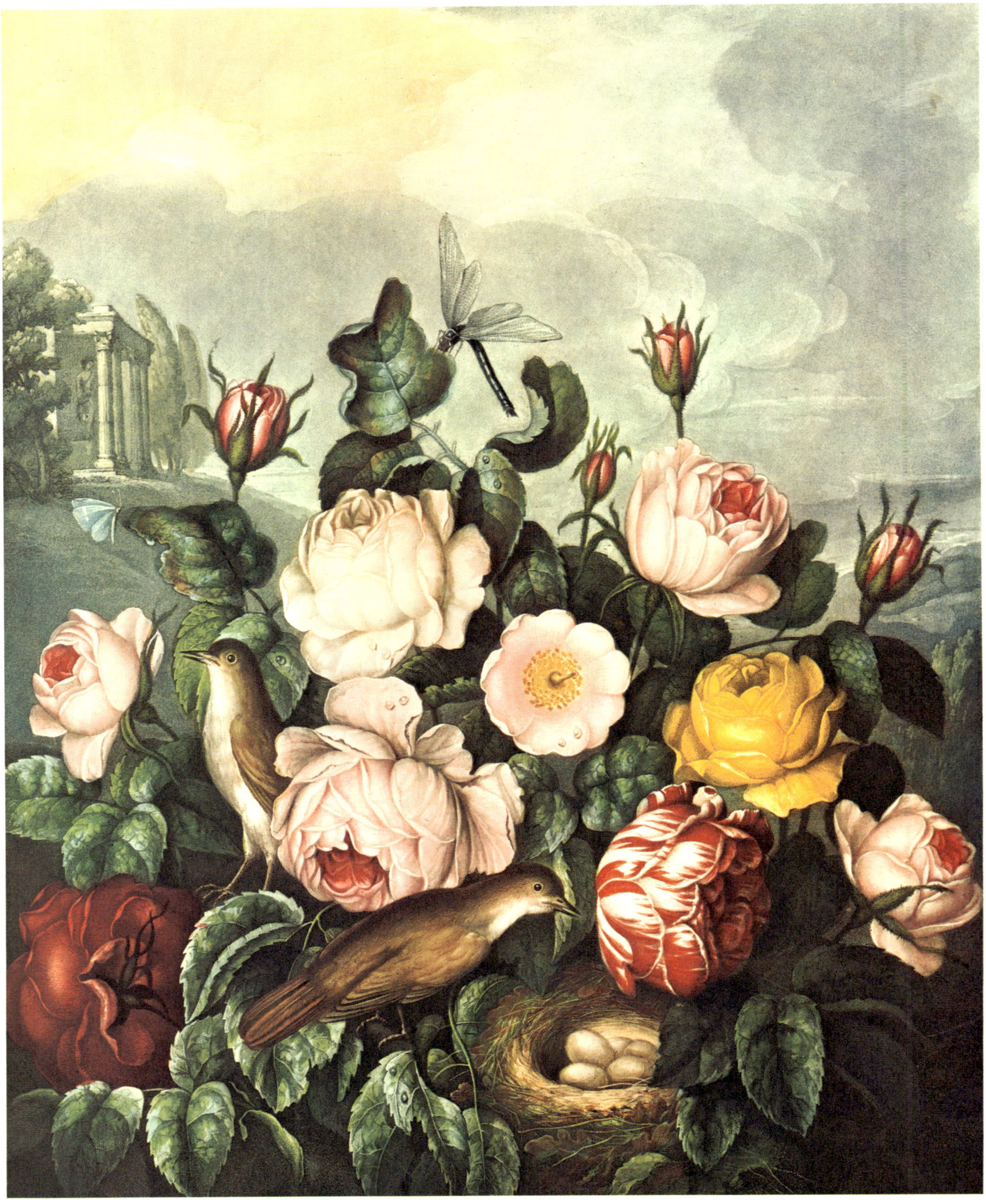

Robert John Thornton M.D. pinxt

Earlom sculpt

Roses

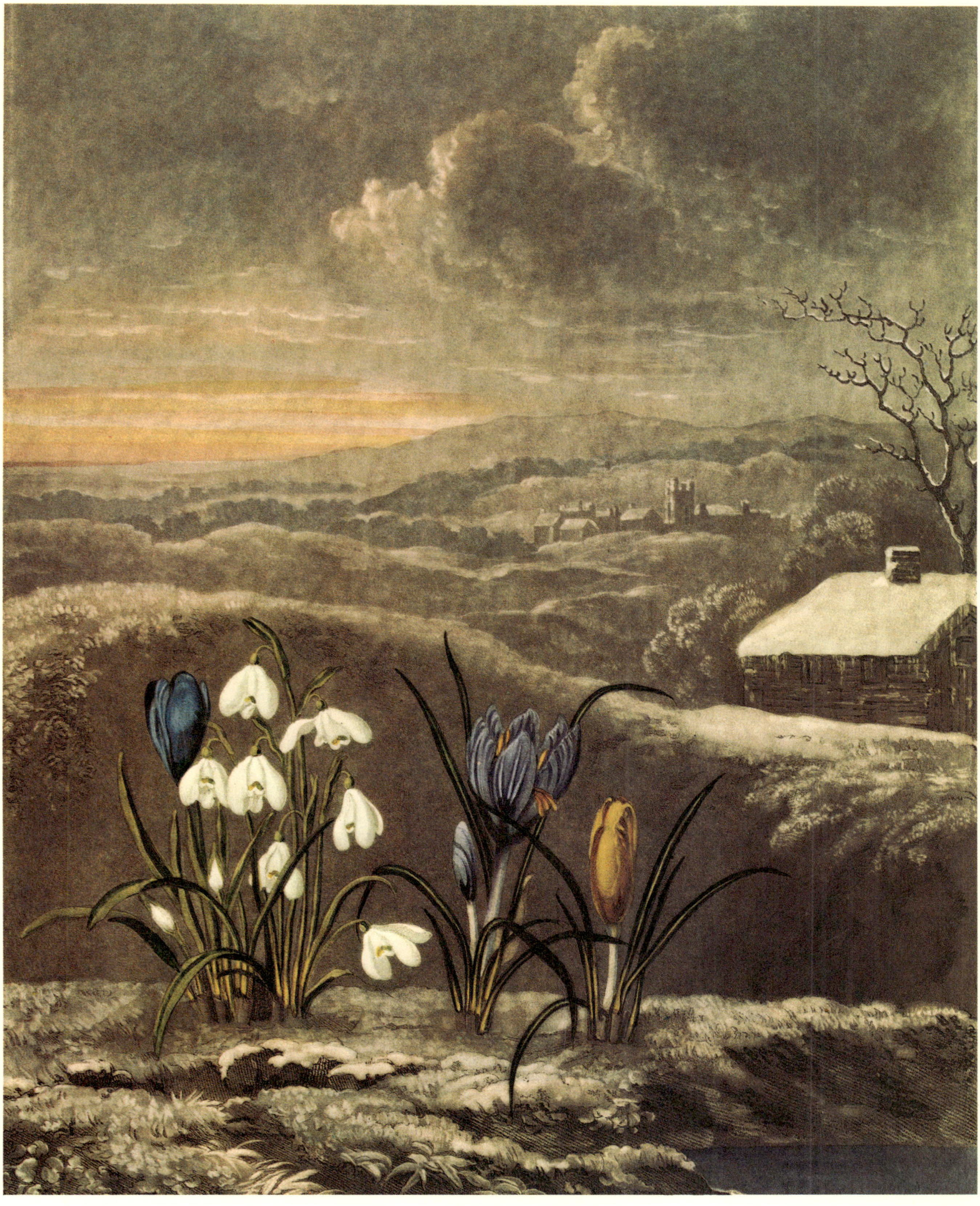

Pether pinx[t] W Ward Engraver to H.R.H the Duke of York, sculp[t]

The Snowdrop

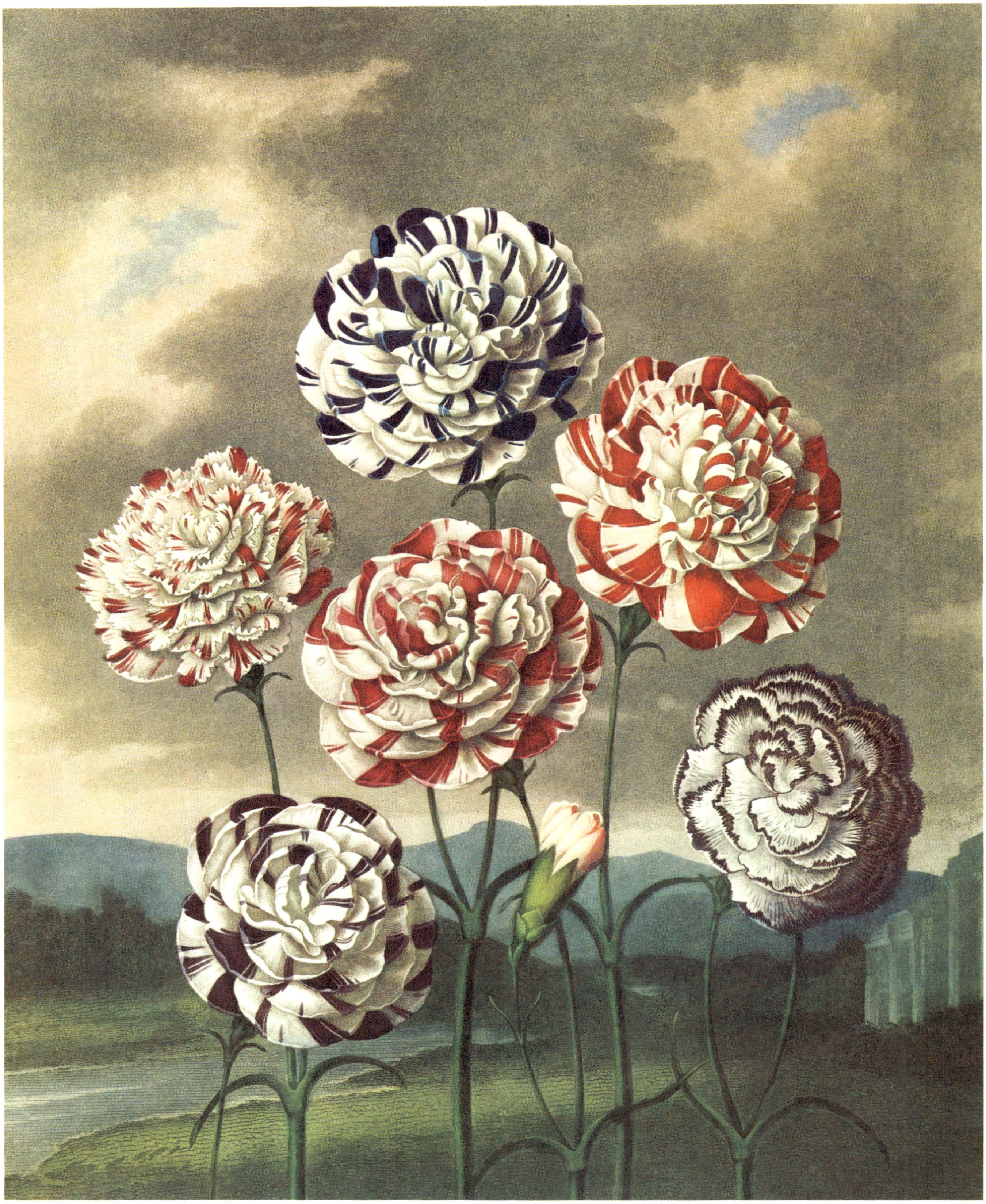

Henderson pinx — Caldwall sculp

A Group of Carnations

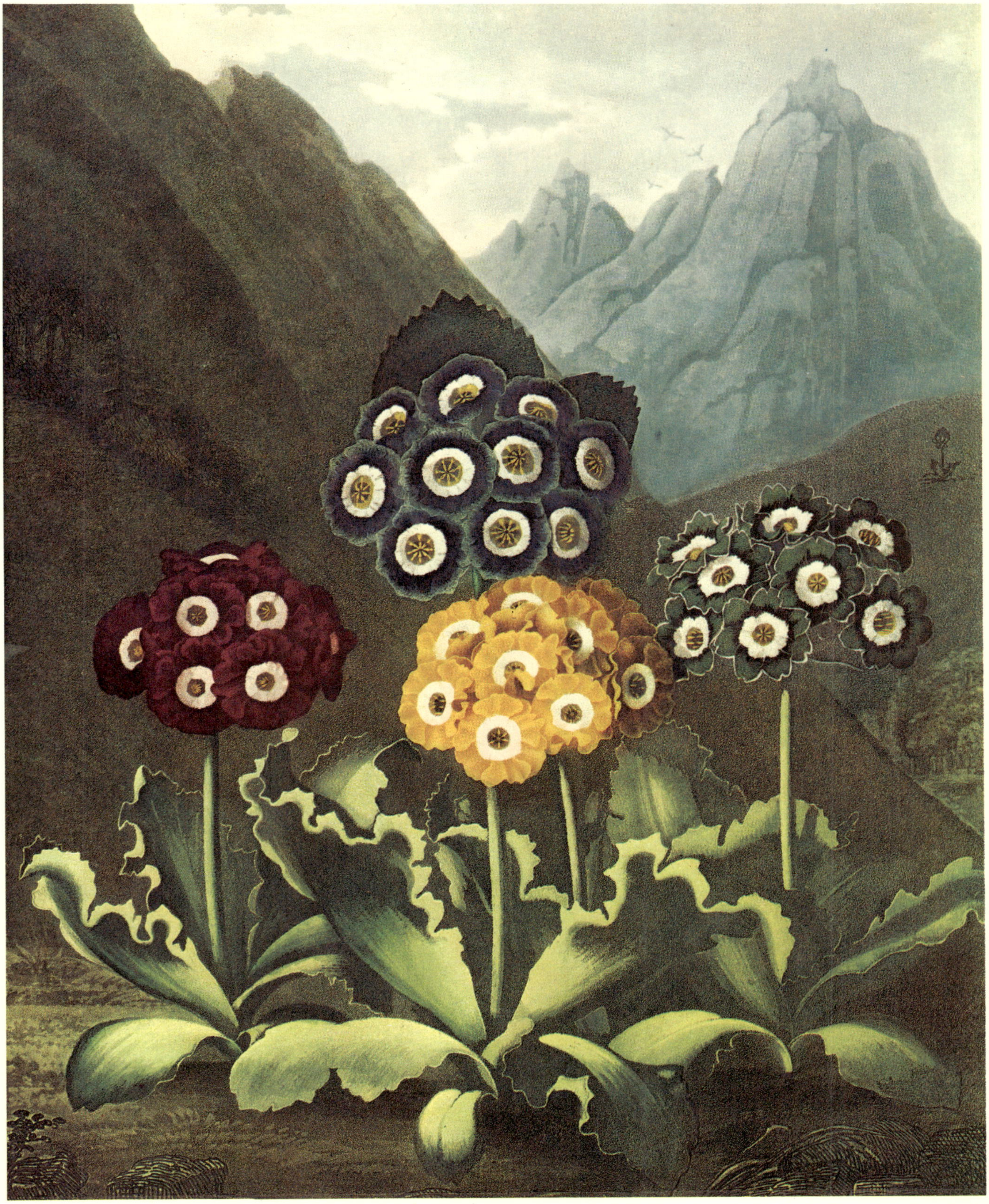

Henderson pinx. Lavis & Hopwood sculp.

A Group of Auriculas.

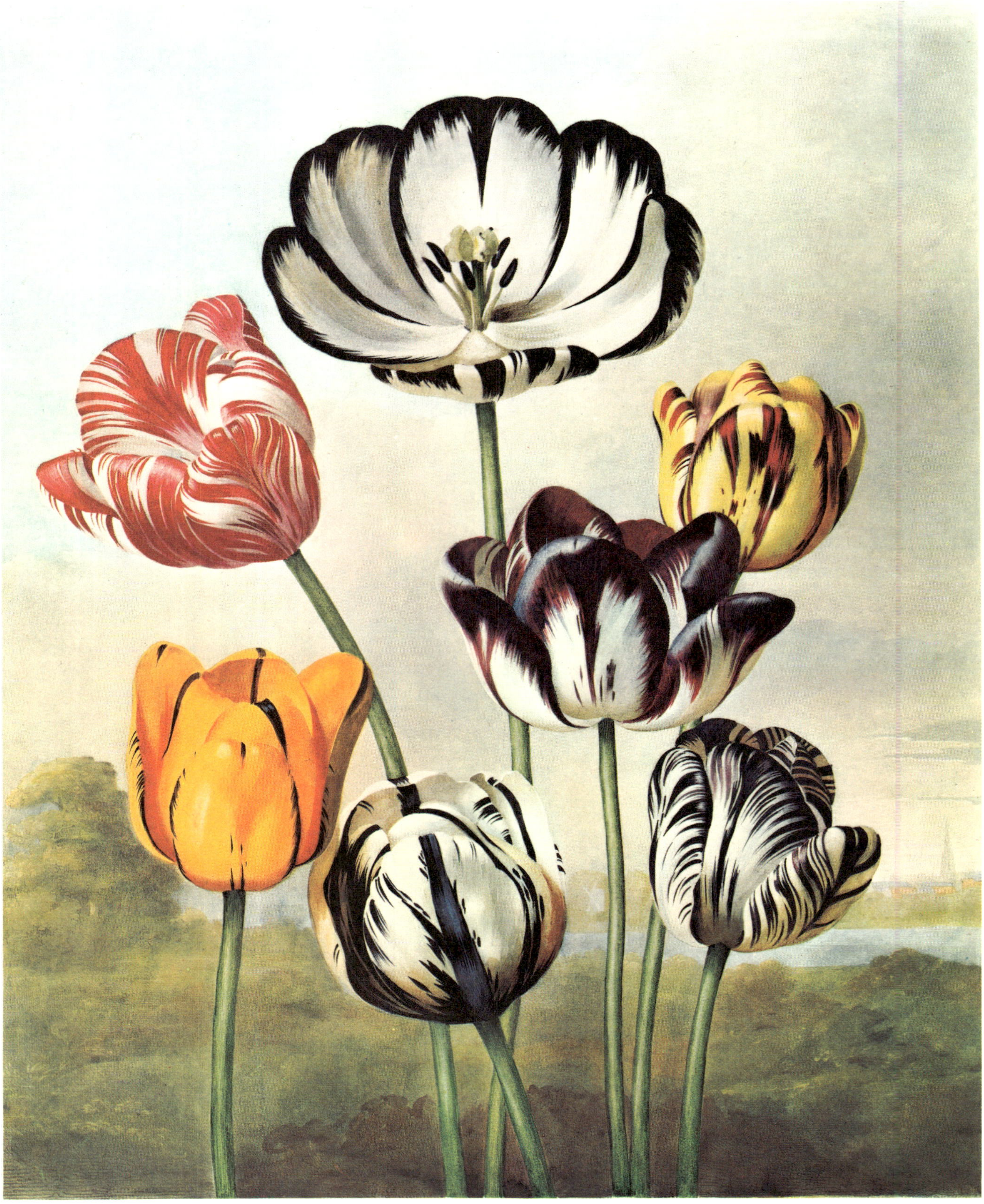

Reinagle pinx.t

Tulips

Earlom sculp.t

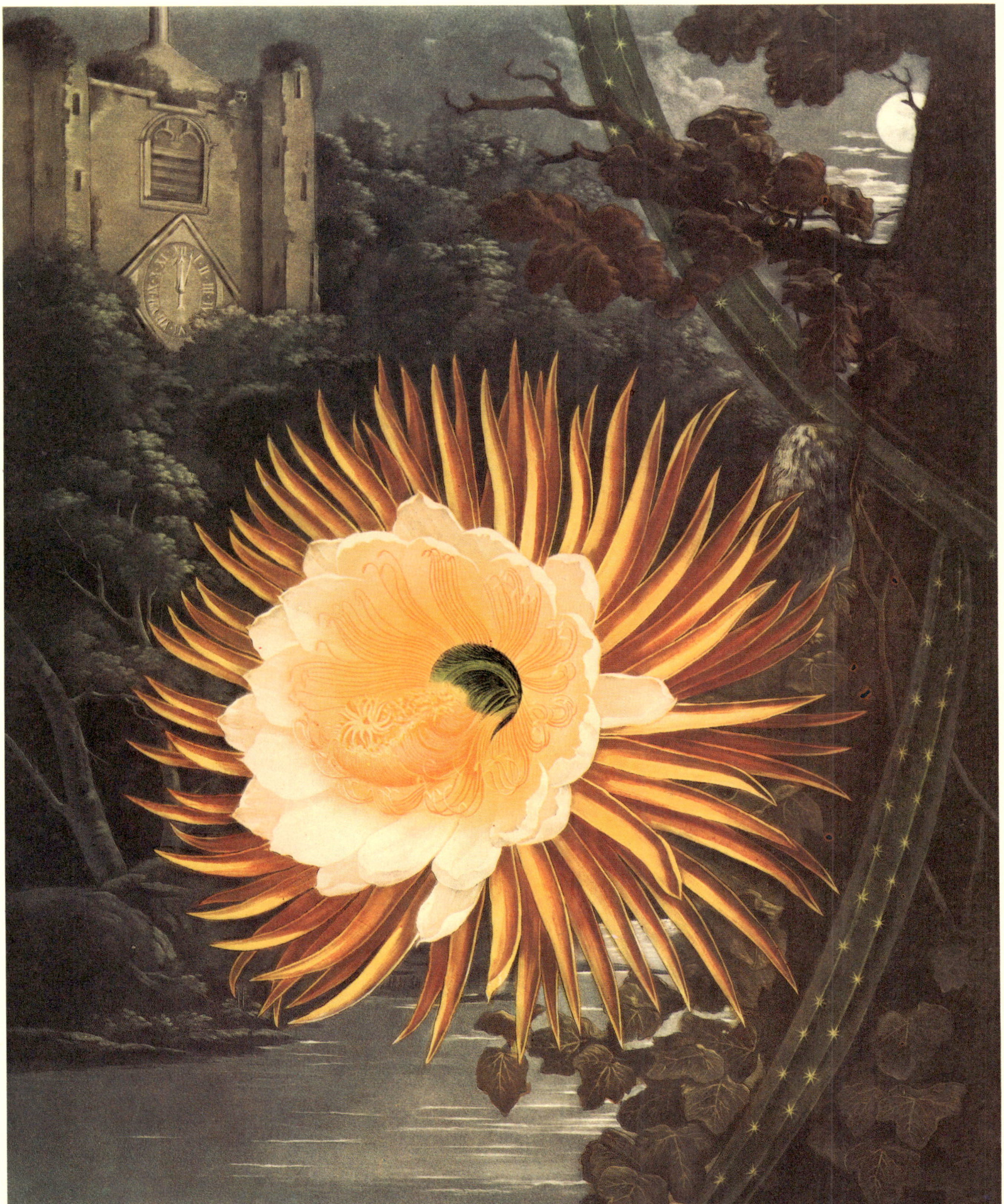

The Flower by Reinagle, Moon-light by Pether.   Dunkarton sculpt

The Night-Blowing Cereus.

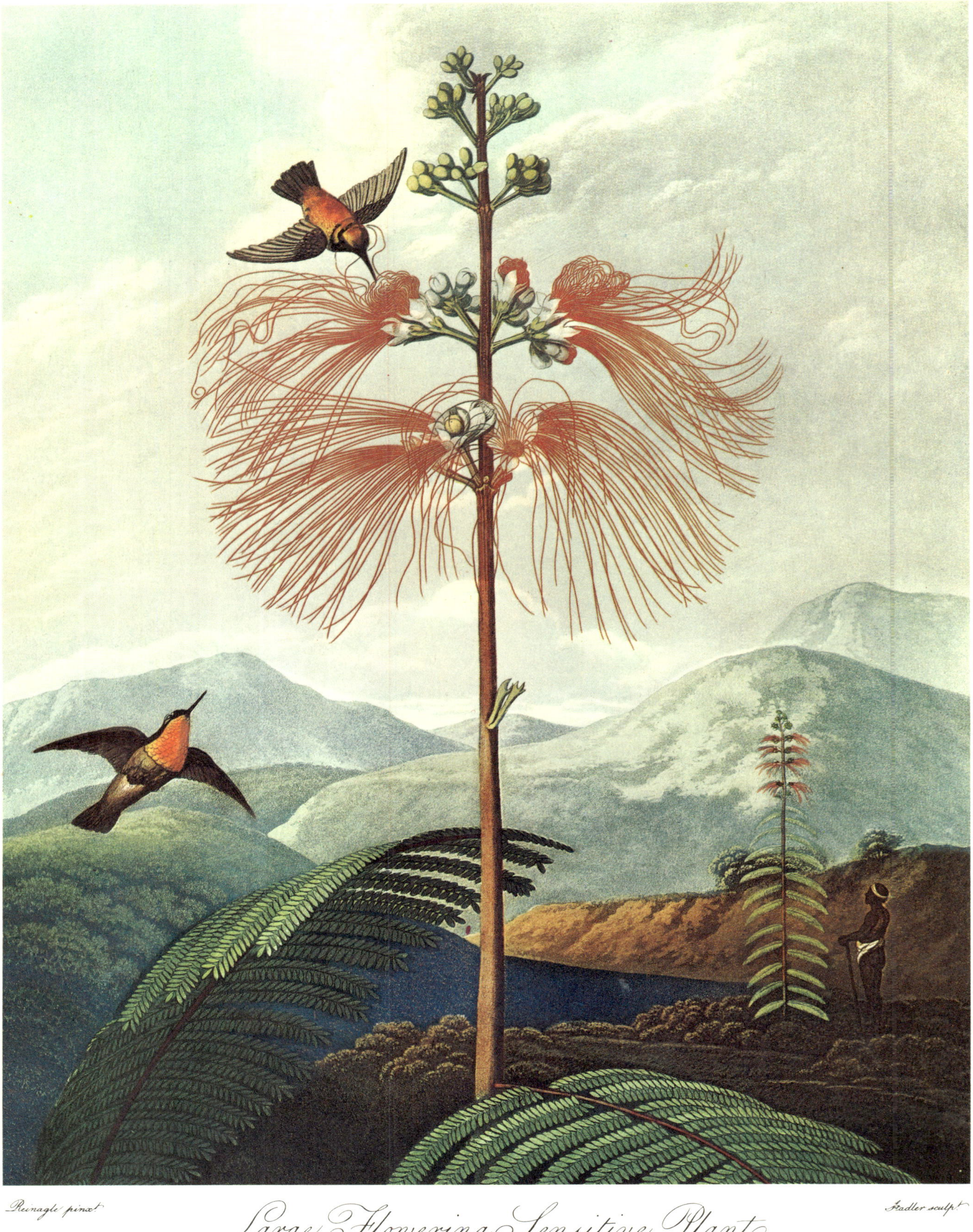

Reinagle pinx^t

Stadler sculp^t

Large Flowering Sensitive Plant

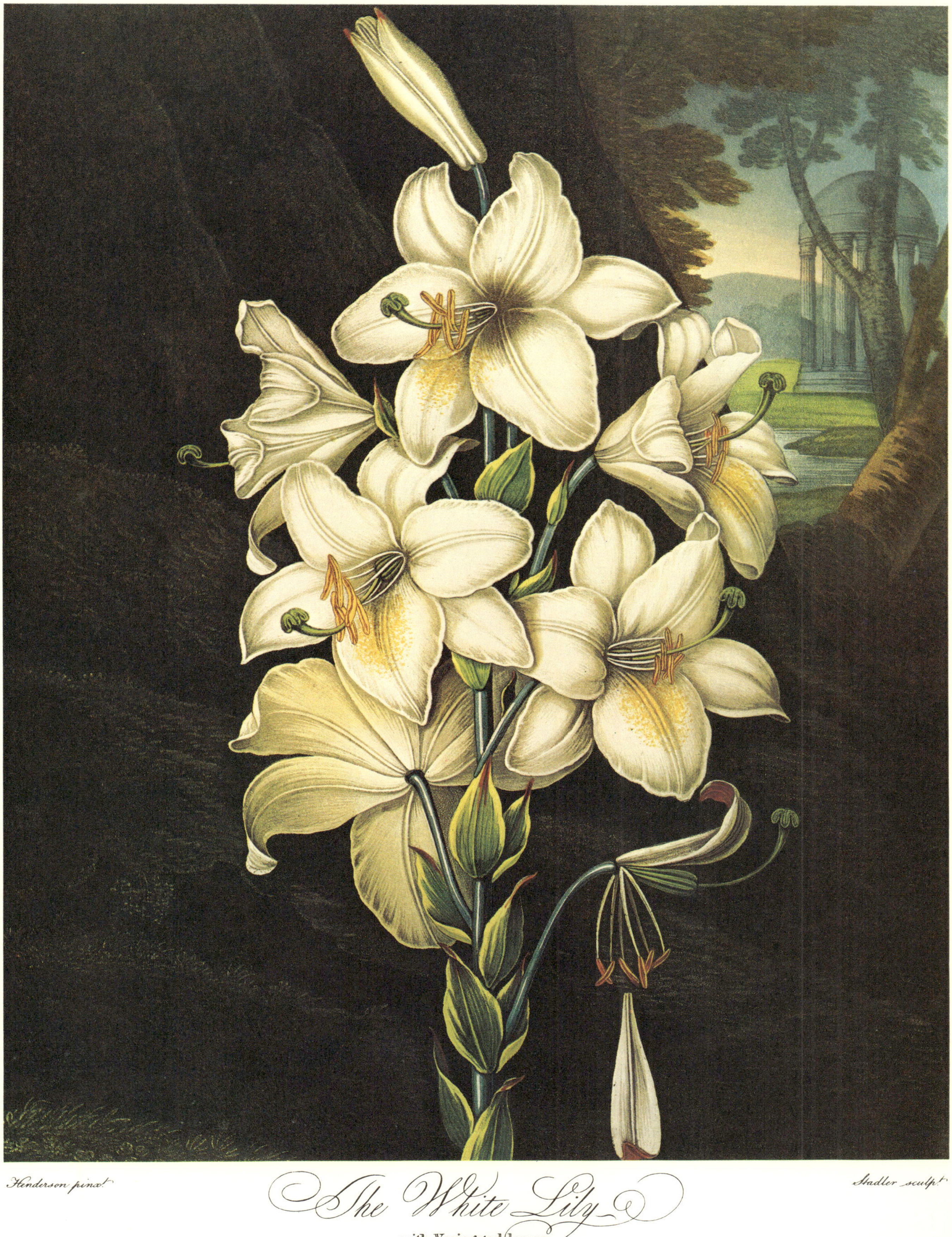

Henderson pinx.t

Stadler sculp.t

The White Lily

with Variegated leaves.

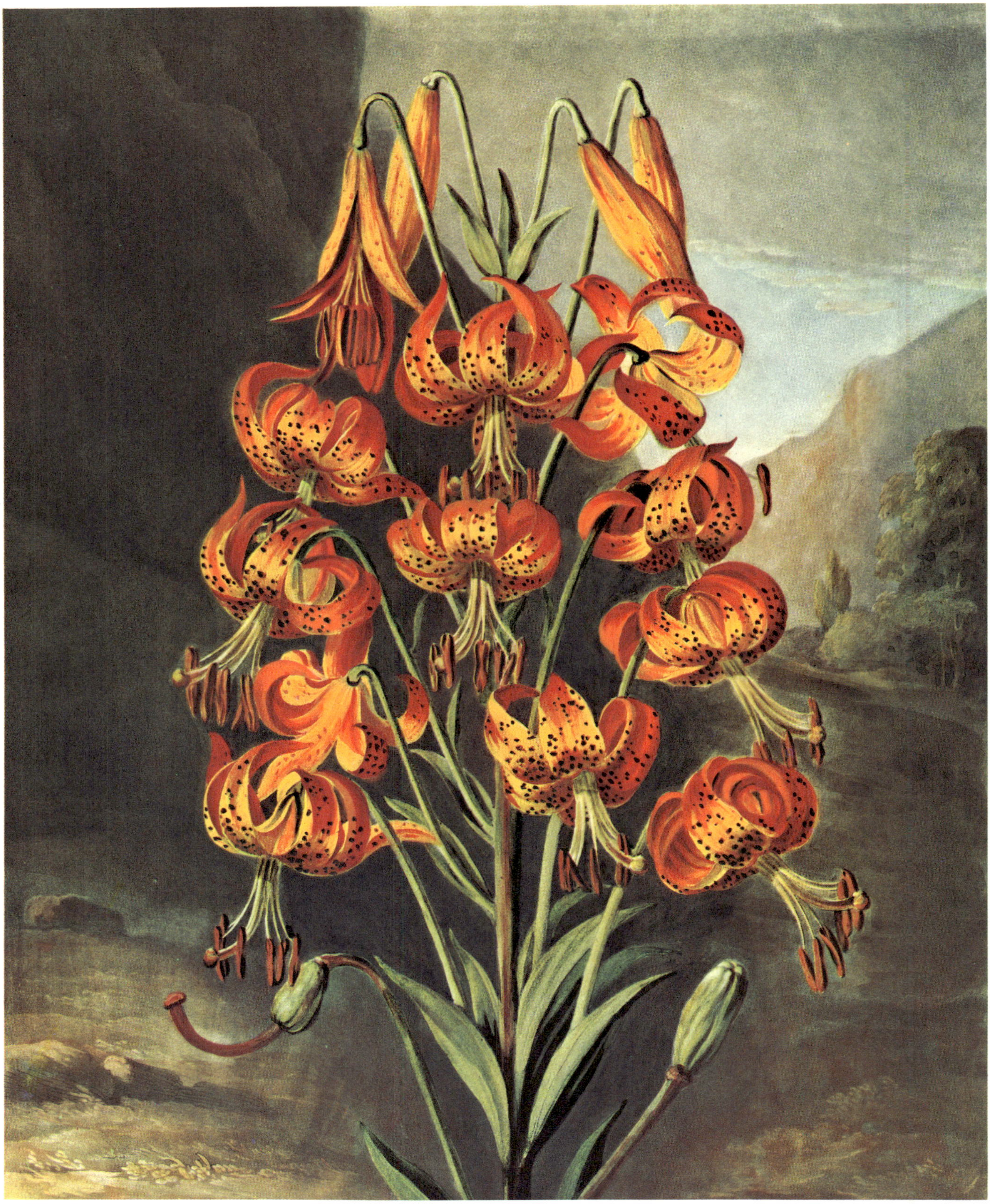

Reinagle pinxt. *Earlom sculpt.*

The Superb Lily

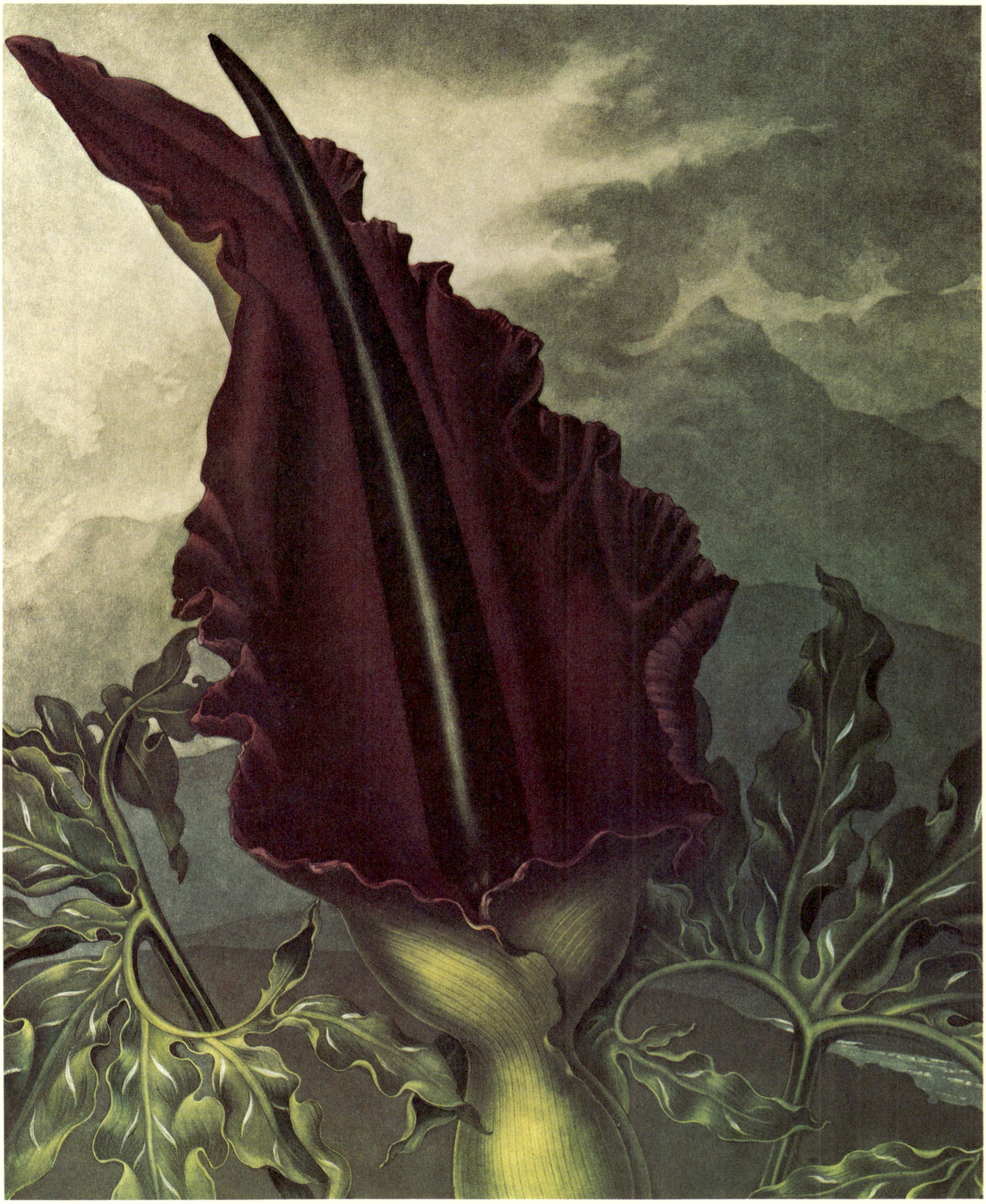

Henderson pinx.t *Ward sculp.t*

The Dragon Arum.

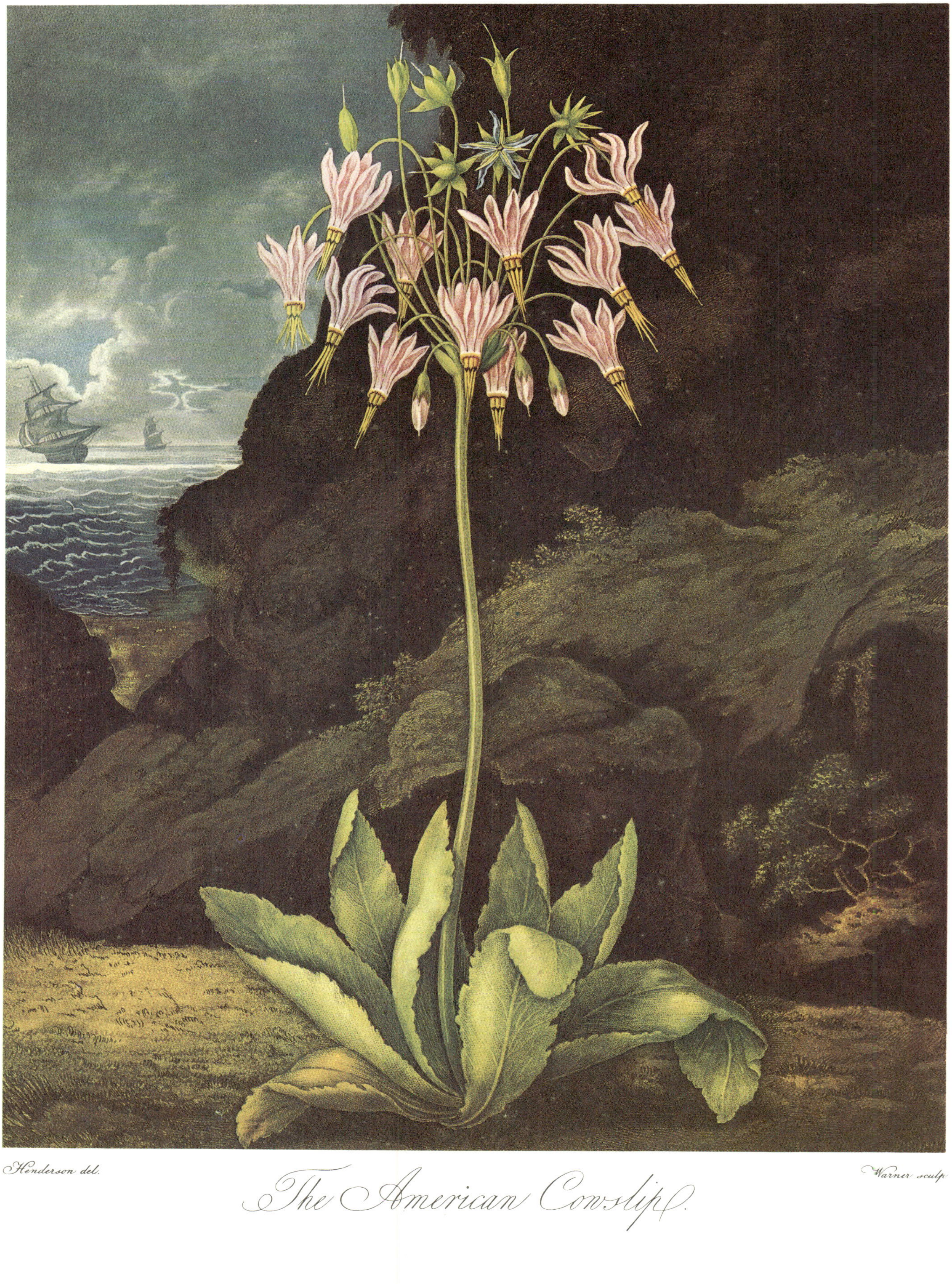

Henderson del. Warner sculp.

The American Cowslip.

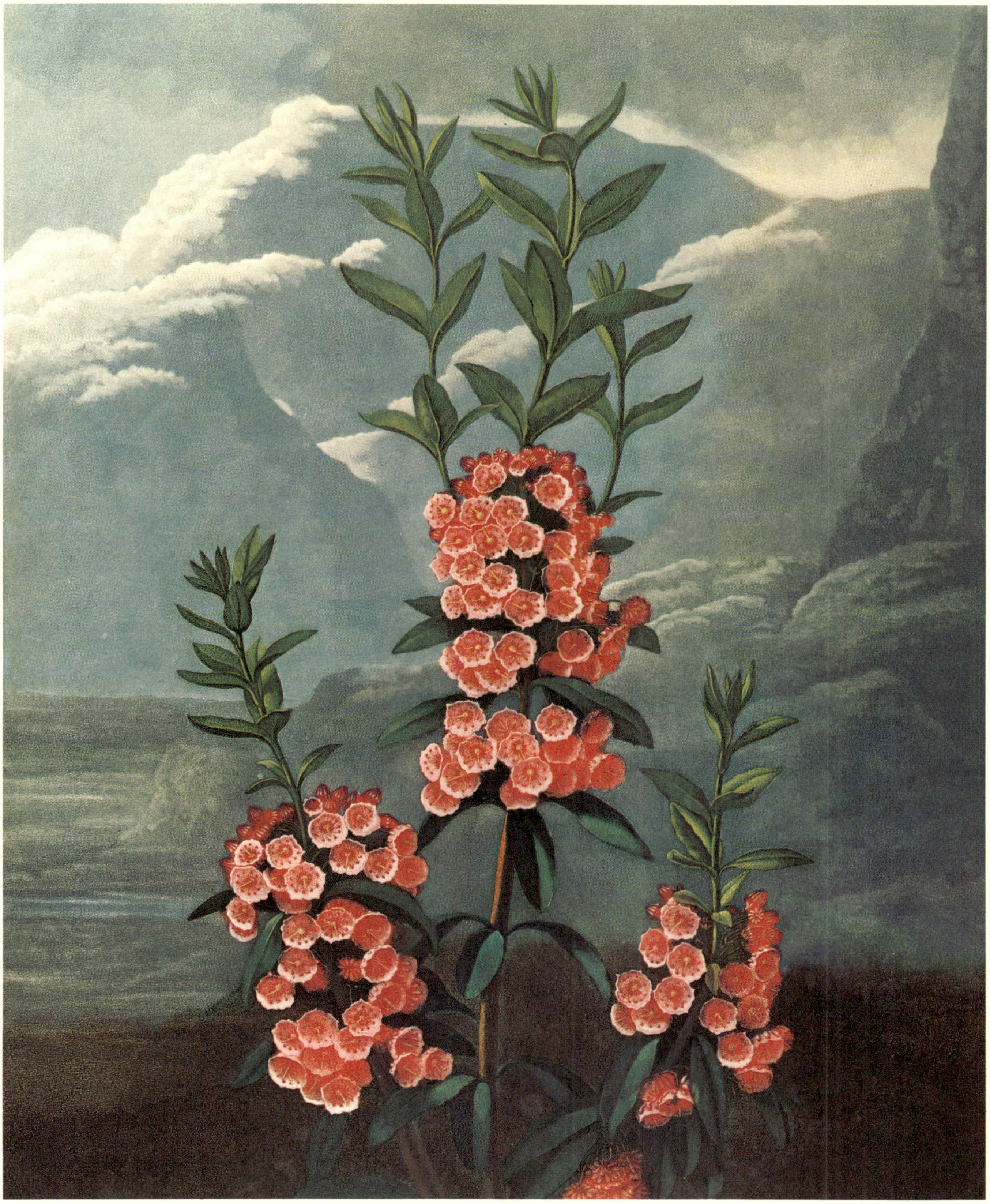

Reinagle A.R. pinx[t] Caldwall sculp[t]

The Narrow-leaved Kalmia.

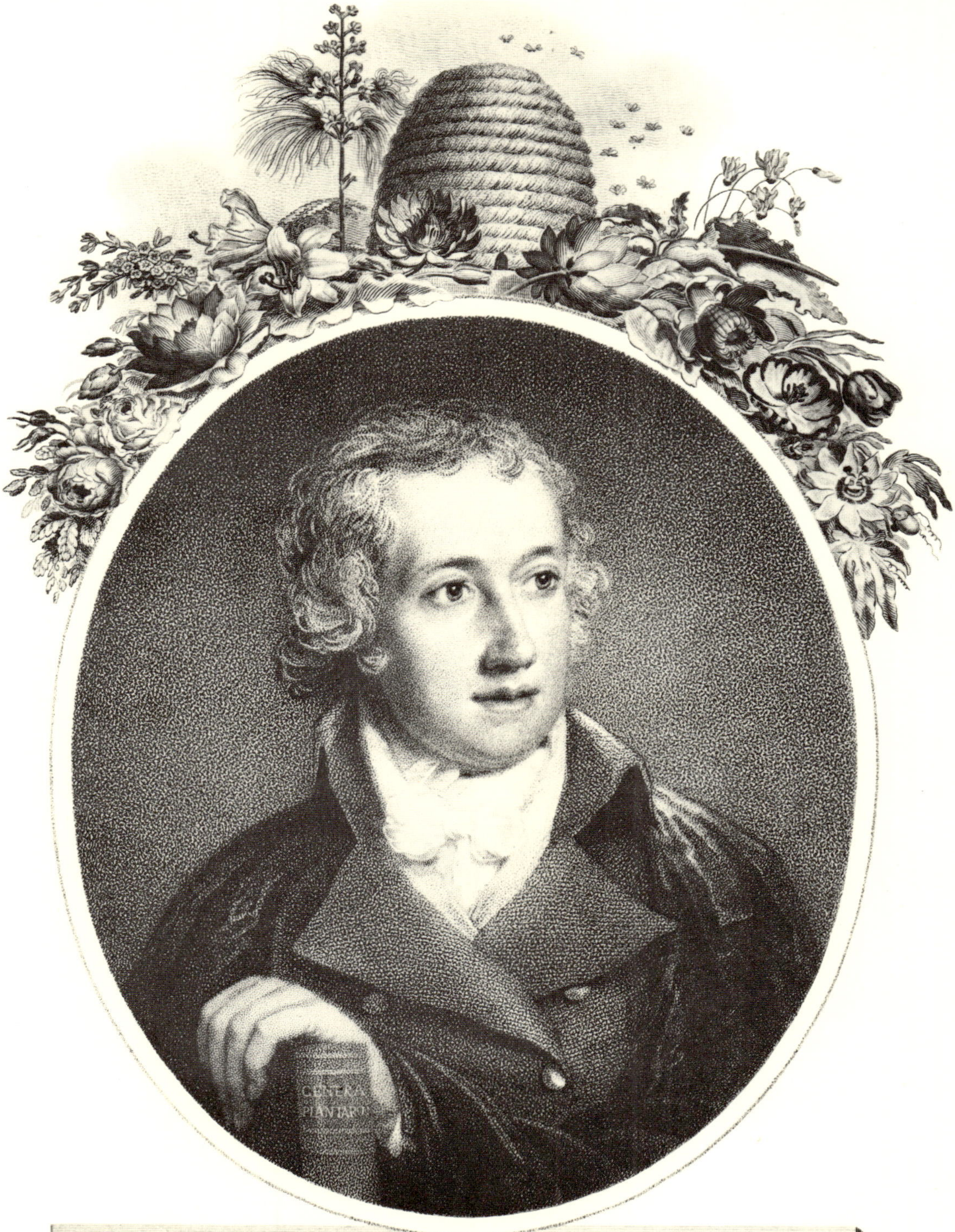

ROBERT JOHN THORNTON, M.D.
Public Lecturer on Medical Botany.

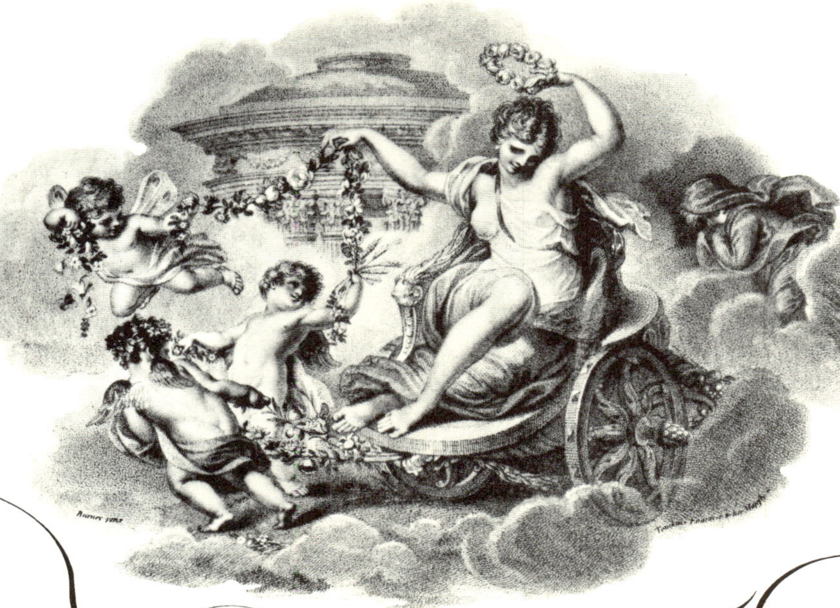

THE Temple of Flora,

OR

GARDEN

OF

Nature,

CONTENTS

Tomkins scr. | *Vincent fec.*

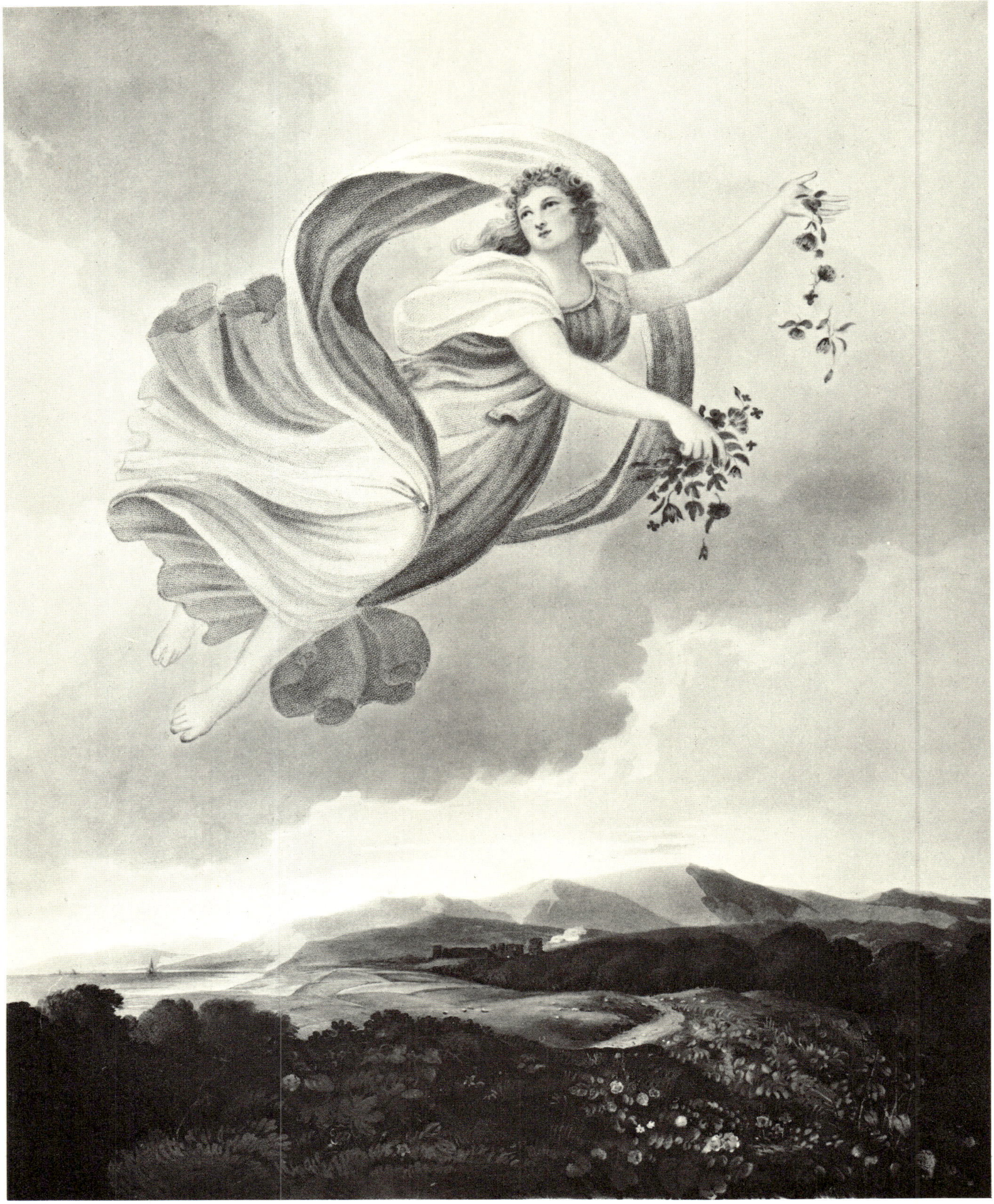

Cosway R.A. pinx. T. Woolnoth sculp

Flora dispensing Her Favours on the Earth

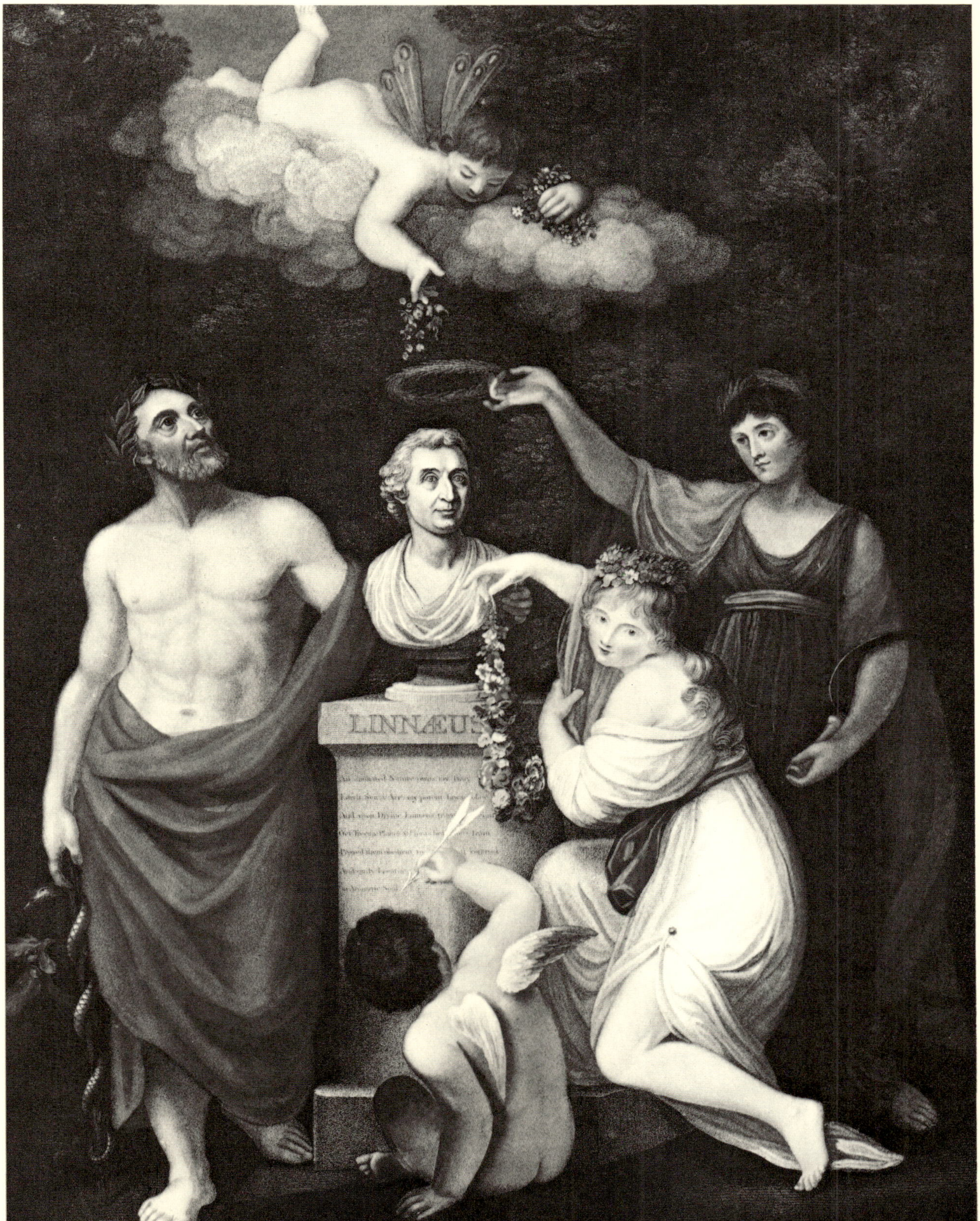

Rufsel R.A. and Opie R.A. pinx — Caldwall sculp

Asculapius, Flora, Ceres and Cupid honouring the Bust of Linnæus.

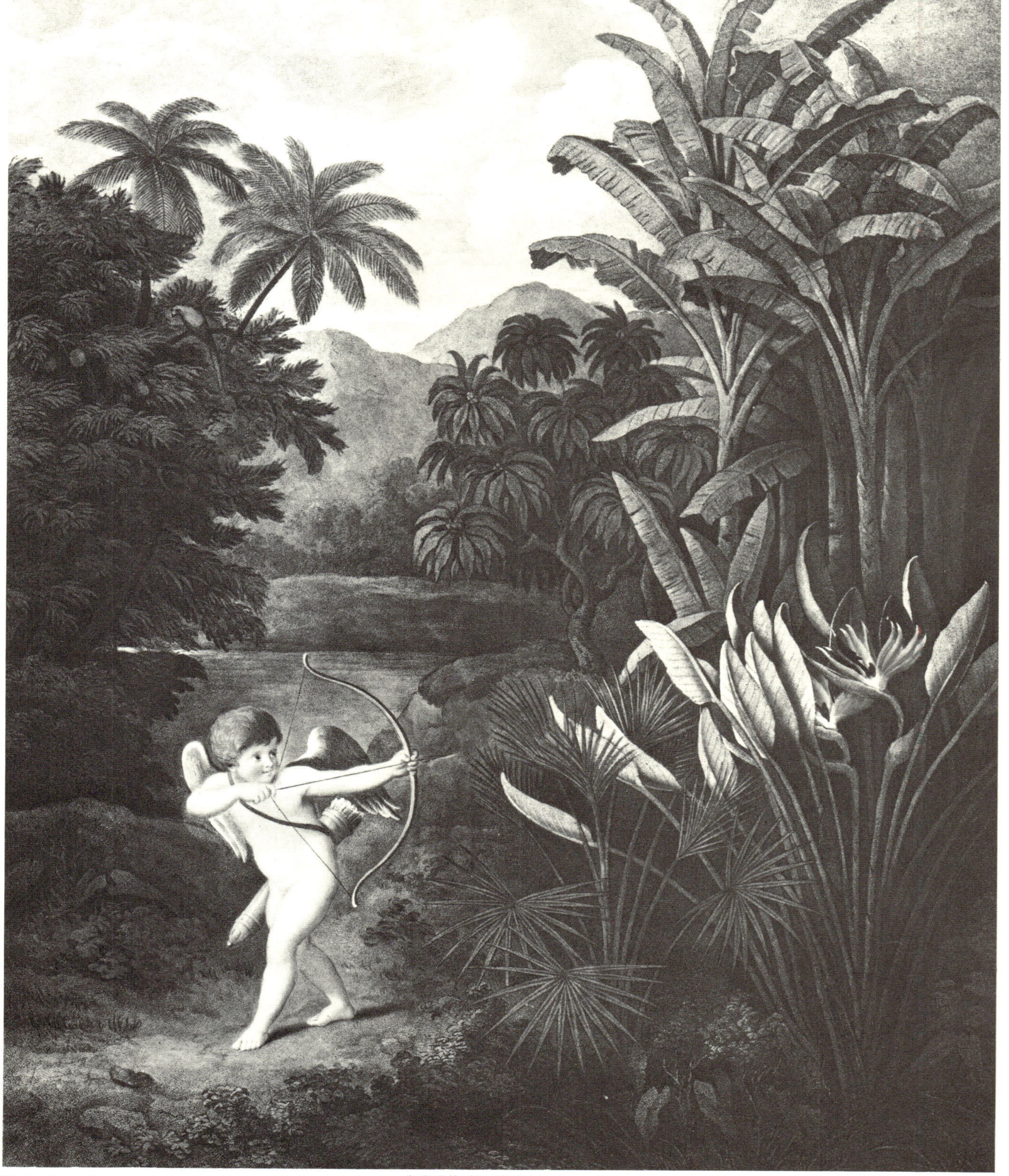

Reinagle Senr. A.R.A. pinx^t. London, Published by D^r. Thornton, June 1, 1803. Burke sculp^t.

And thou, divine LINNÆUS! trac'd my Reign,
O'er Trees, and Plants, and Flora's beauteous Train,
Prov'd them obedient to my soft Controul,
And gaily breathe the aromatic soul.

Charlotte Lennox

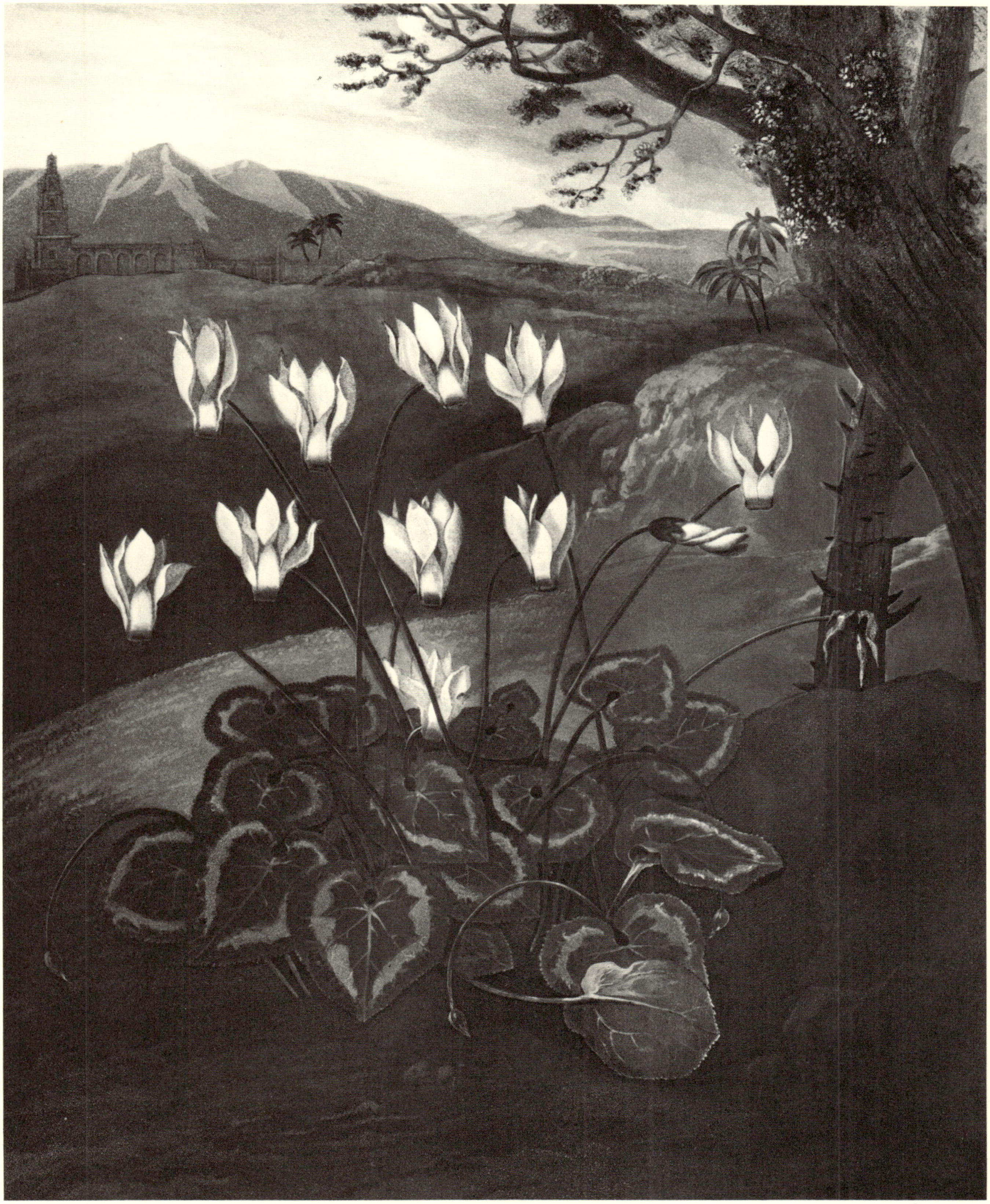

Pether pinxt. Elmes sculpt.

The Persian Cyclamen

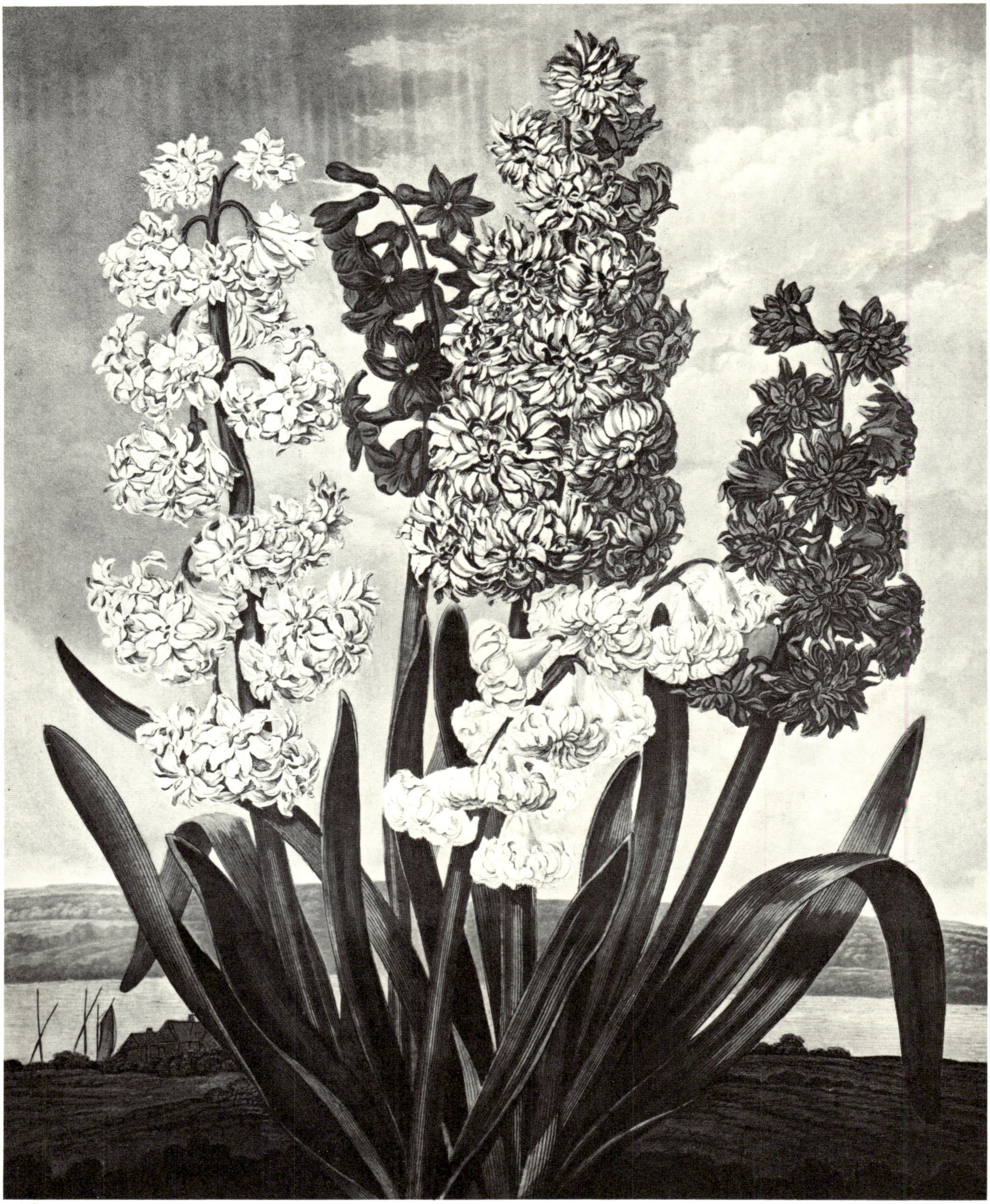

S Edwards pinx^t Warner sculp^t

Hyacinths.

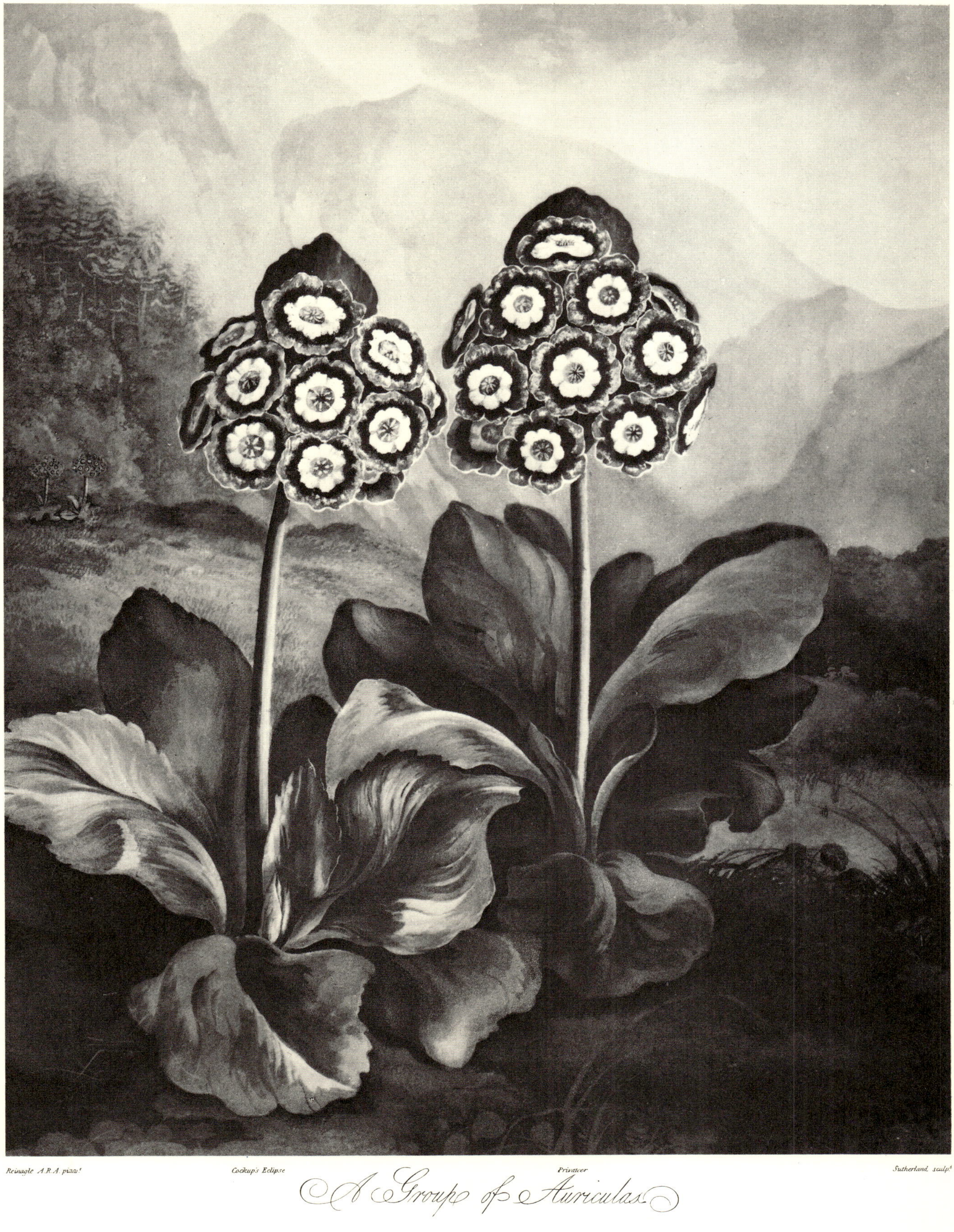

Reinagle A.R.A. pinx.t *Cockup's Eclipse* *Privateer* *Sutherland sculp.t*

A Group of Auriculas

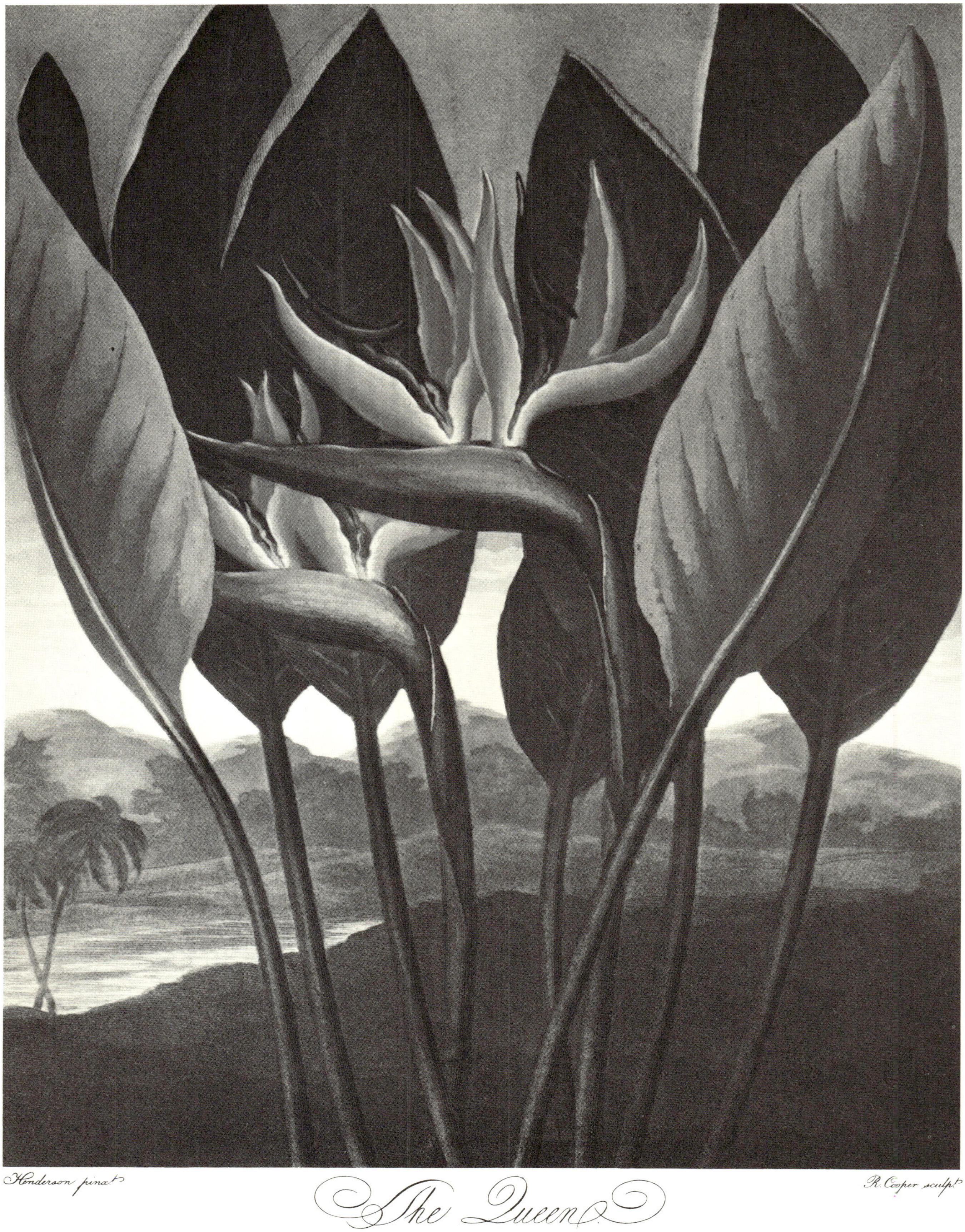

Henderson pinxt — R. Cooper sculpt

The Queen.

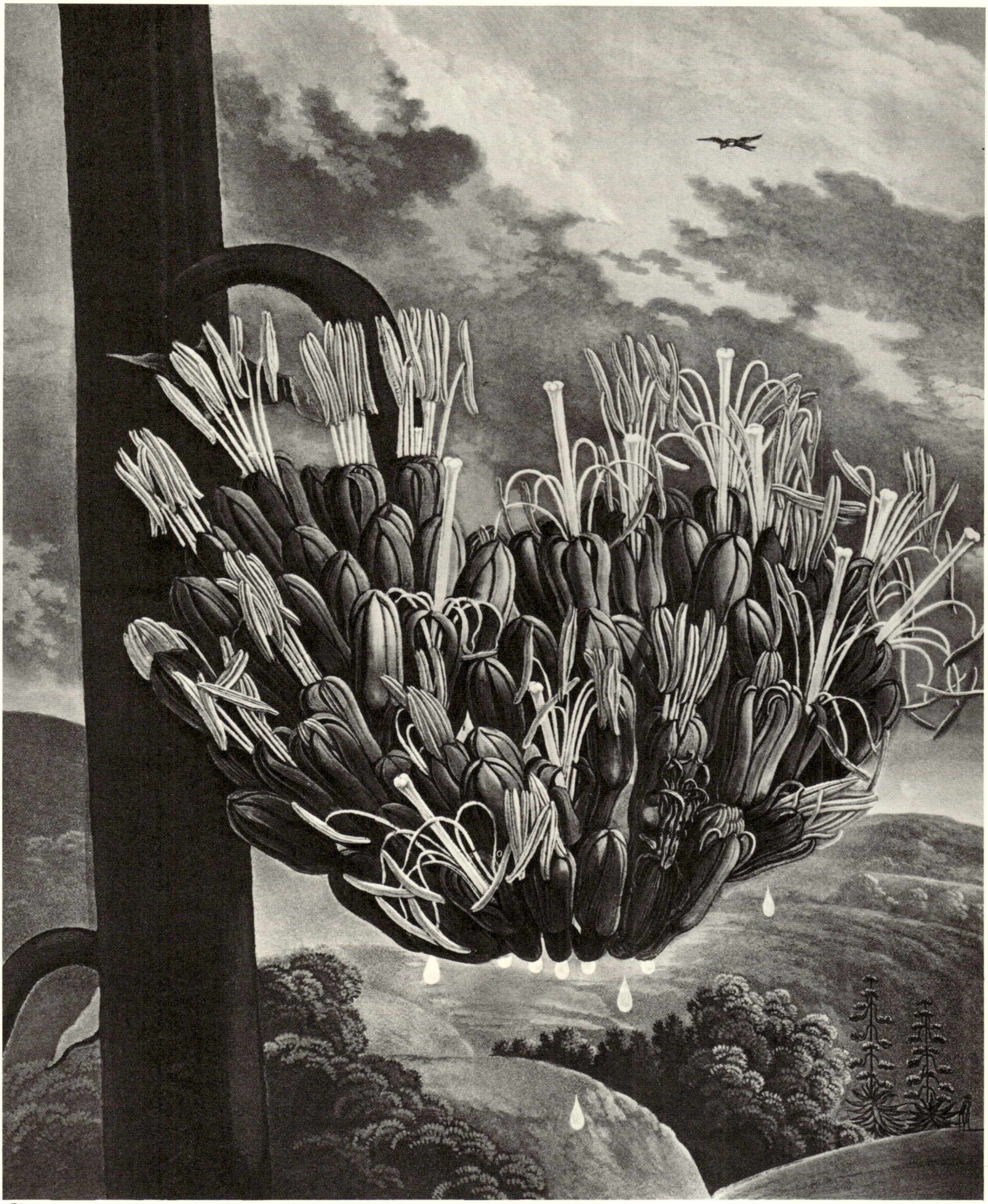

Reinagle pinx^t

Medland sculp^t

The Aloe

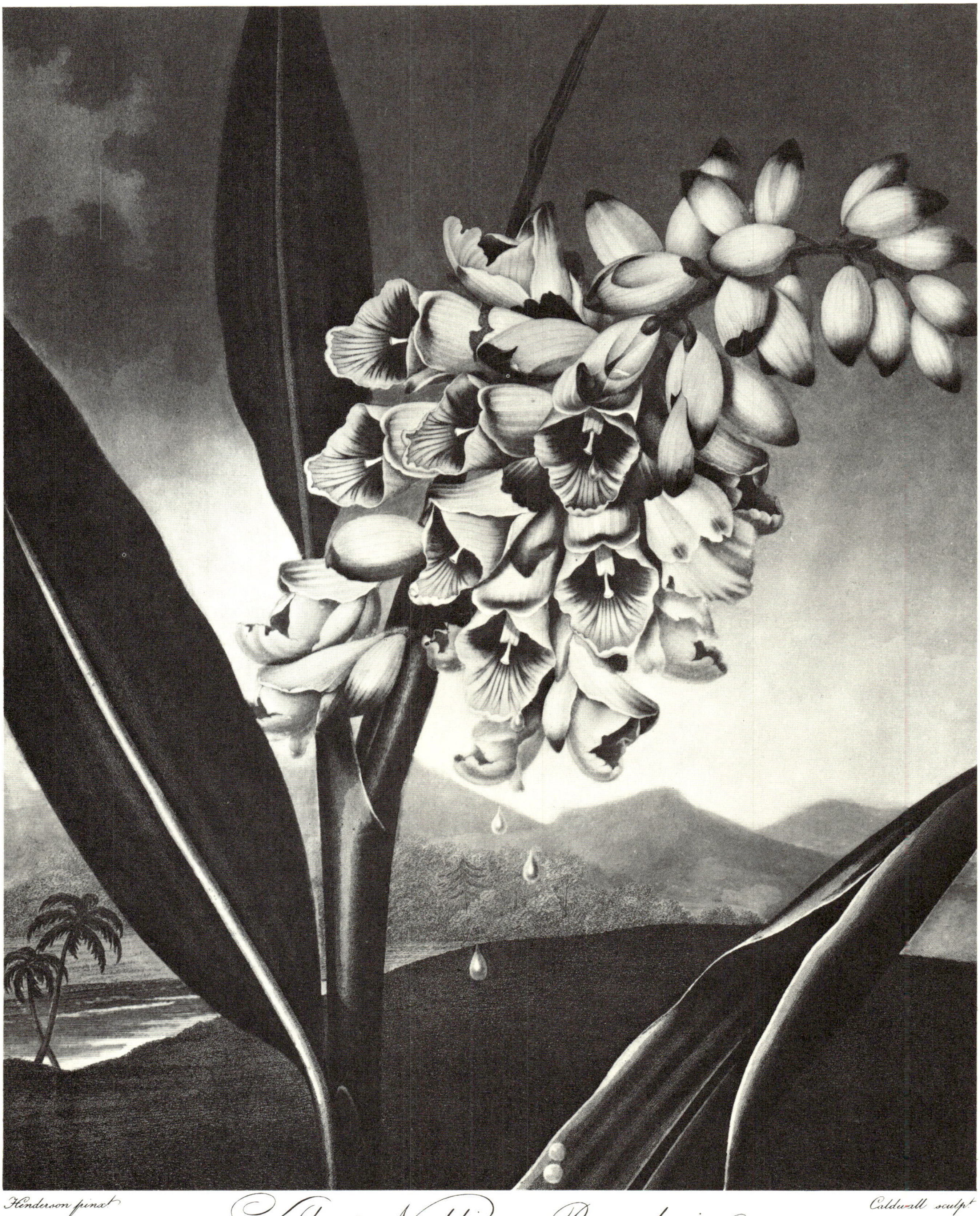

Henderson pinxt

Caldwell sculpt

The Nodding Renealmia

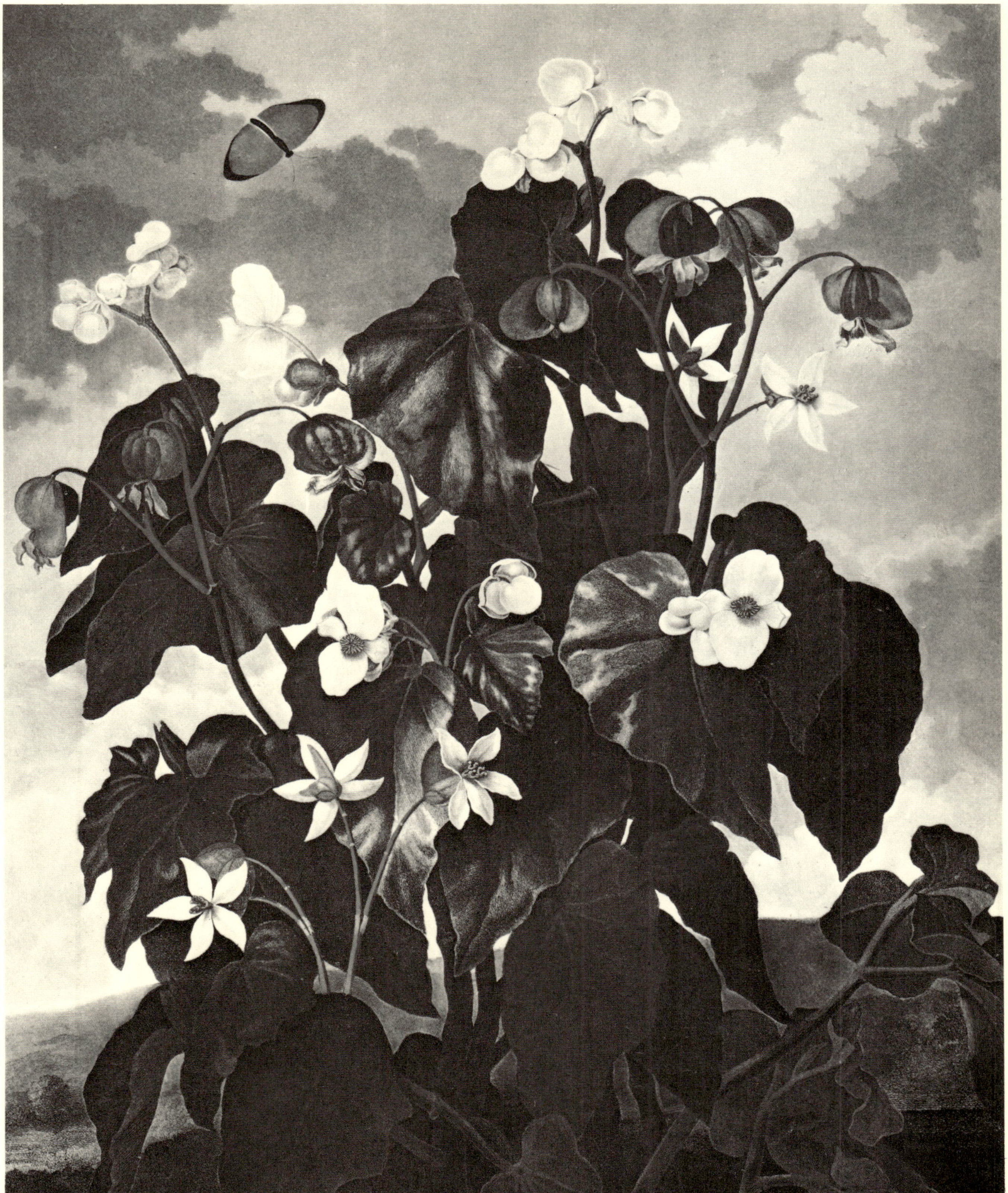

Reinagle pinxt. Caldwall sculpt.

The oblique-leaved Begonia.

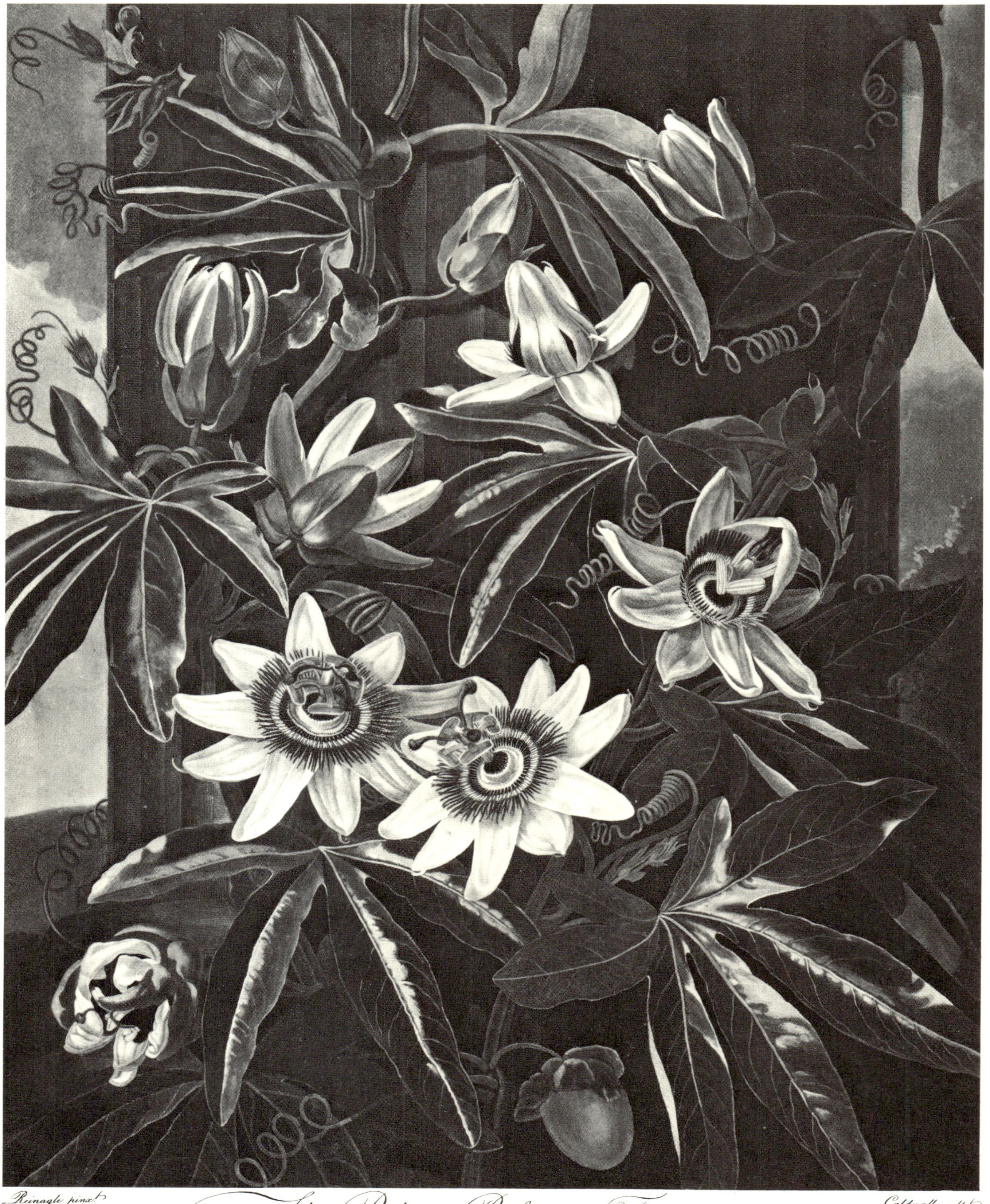

Reinagle pinx.t *Caldwall sculp.t*

The Blue Passion Flower.

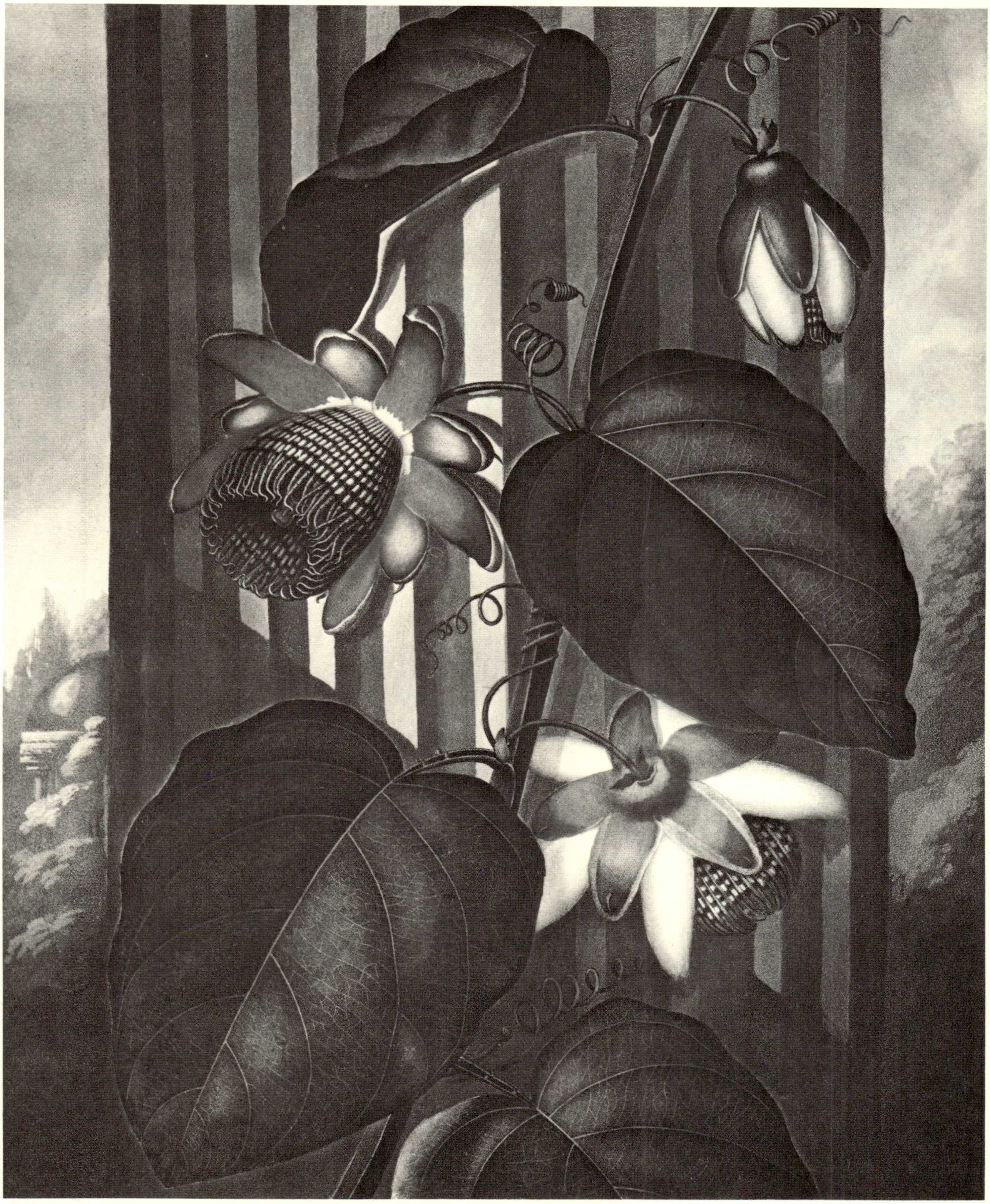

Henderson delt Warner sculpt

The Winged Passion-Flower.

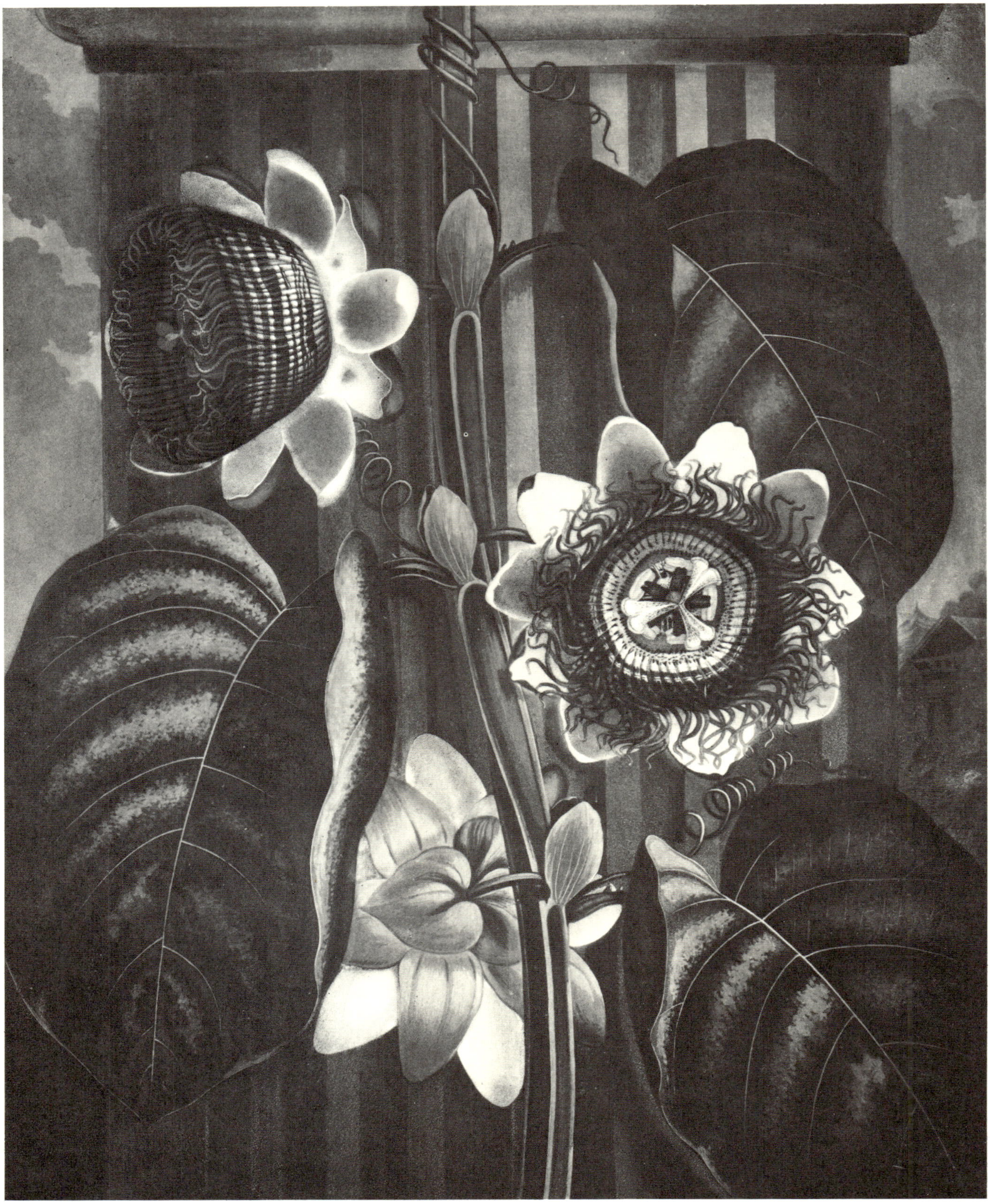

Henderson pinxt

Hopwood sculpt

The Quadrangular Passion-flower.

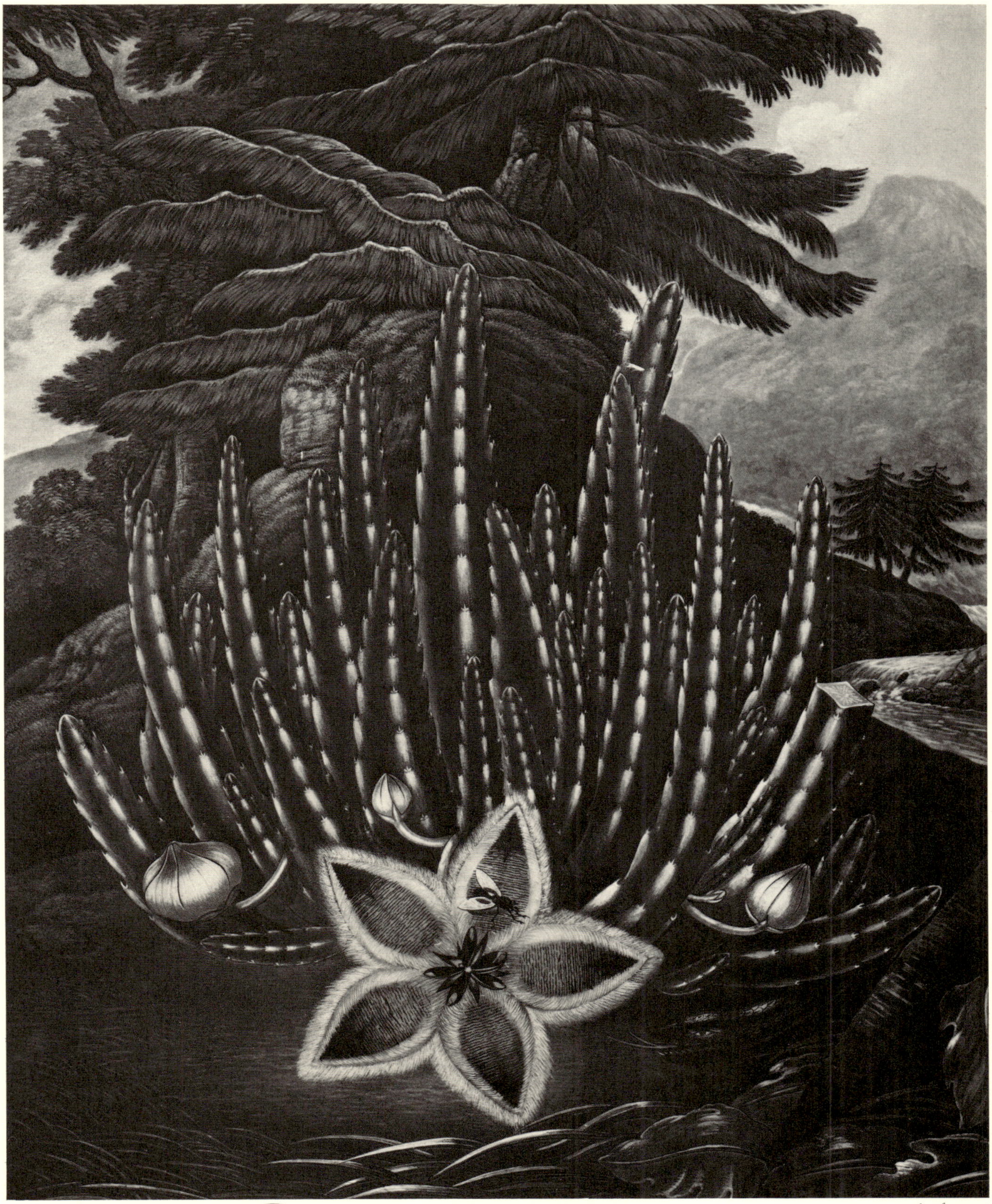

Henderson pinx.t   Stadler sculp.t

The Maggot-bearing Stapelia

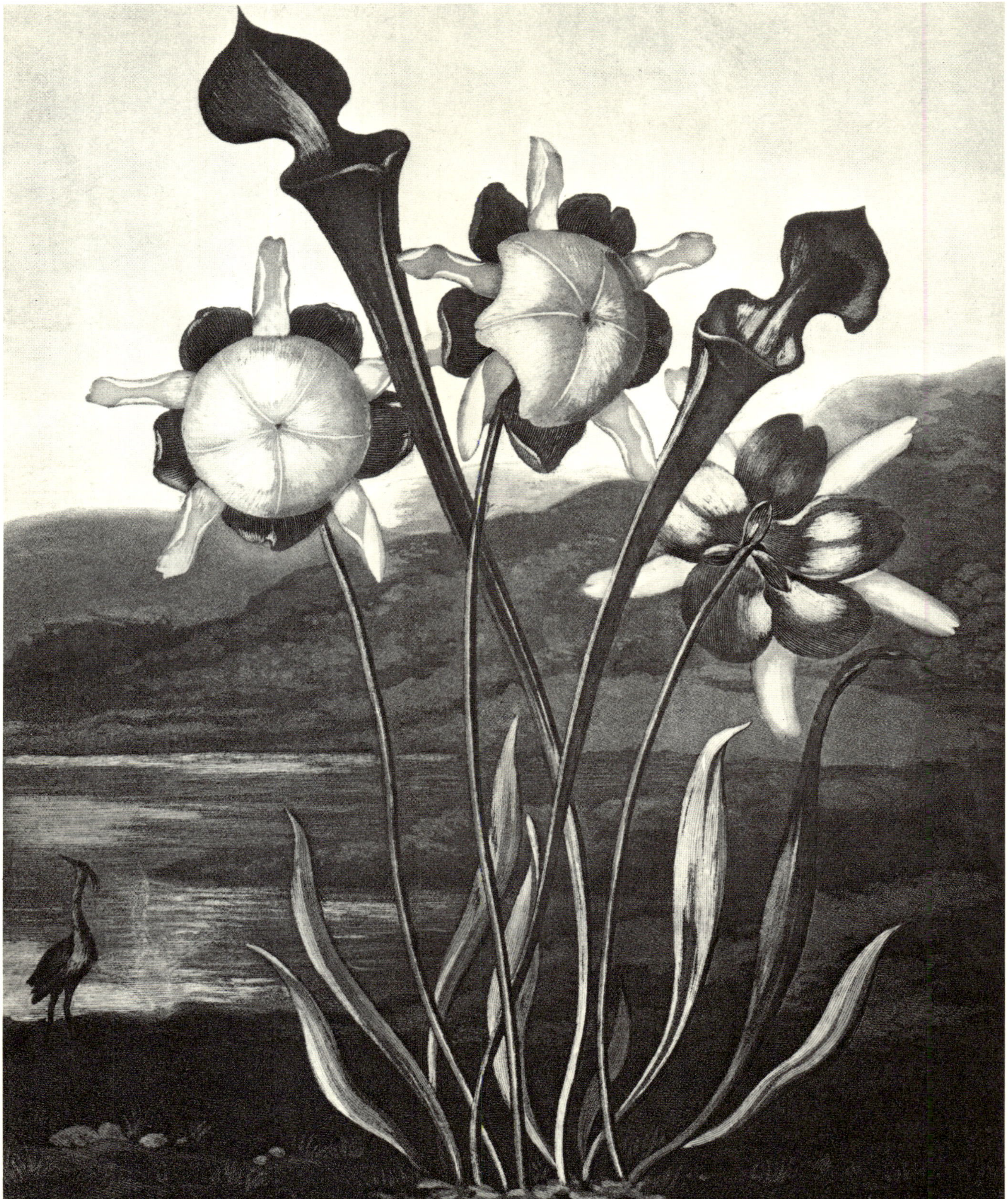

Reinagle Sen.r A pinx.t

R. Cooper sculp.t

Pitcher Plant.

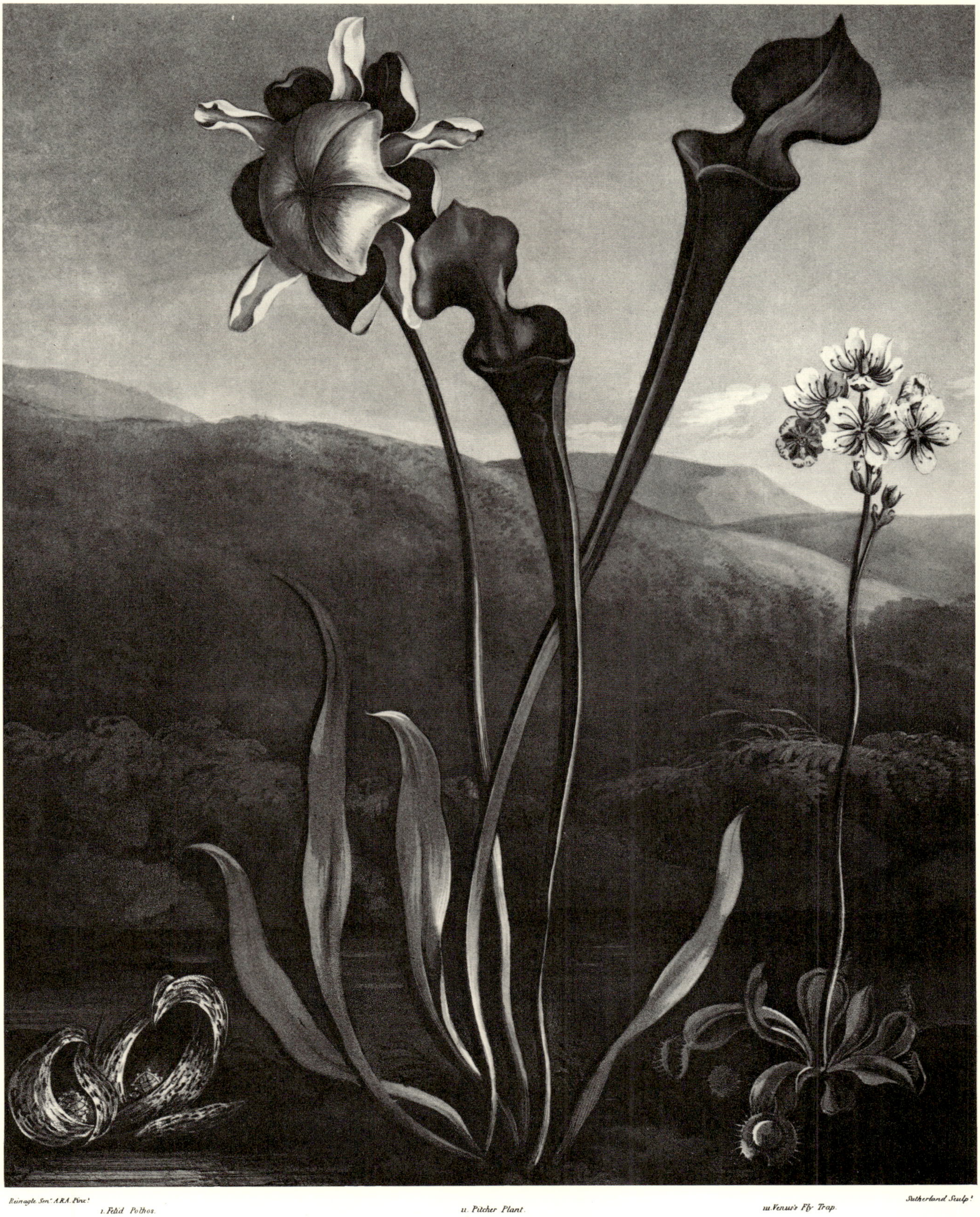

Reinagle Senr A.R.A. Pinx.^t — Sutherland Sculp.^t

i. Fetid Pothos. ii. Pitcher Plant. iii. Venus's Fly Trap.

American Bog-Plants.

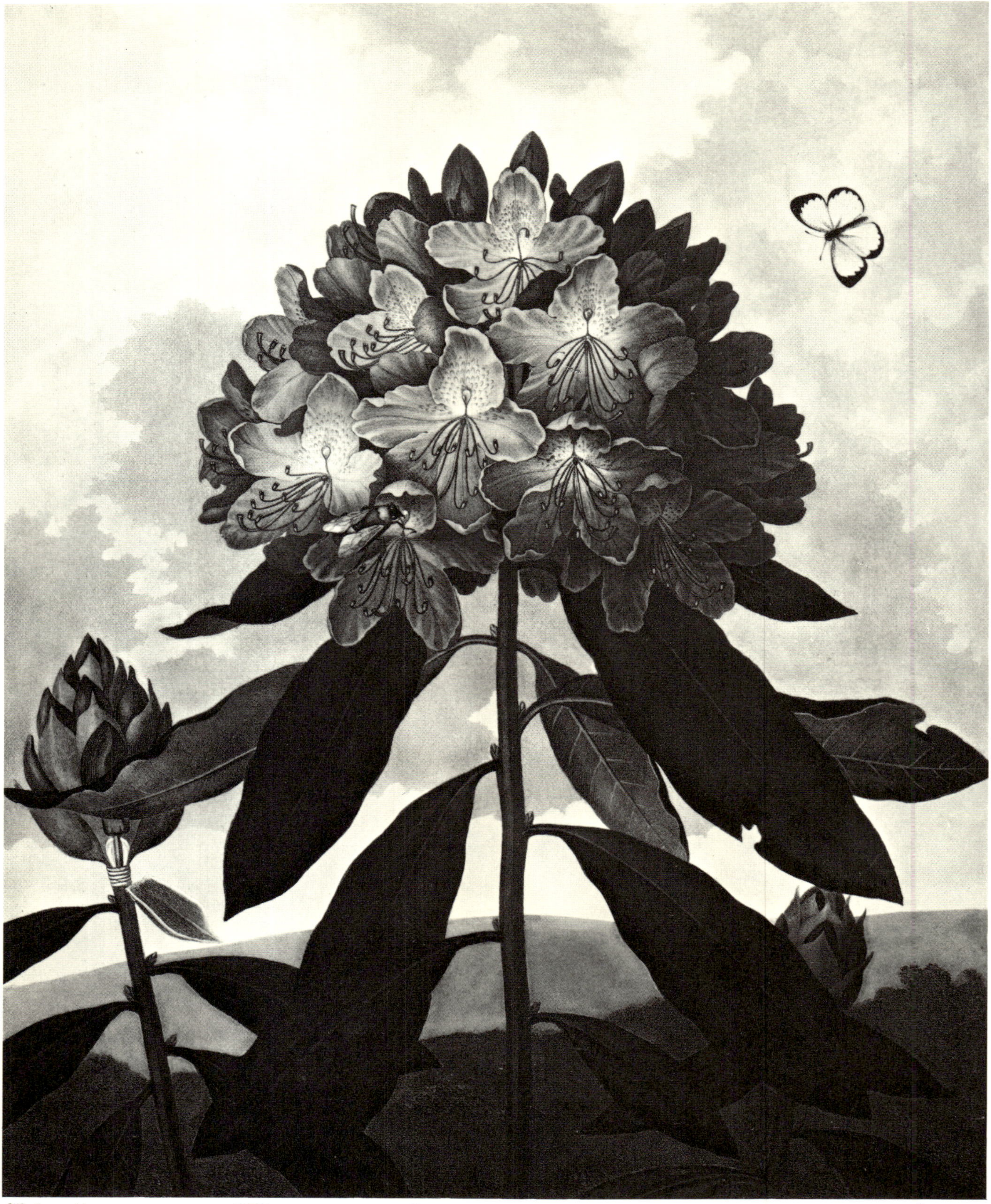

Henderson del. Caldwall sculp.

The Pontic Rhododendron.

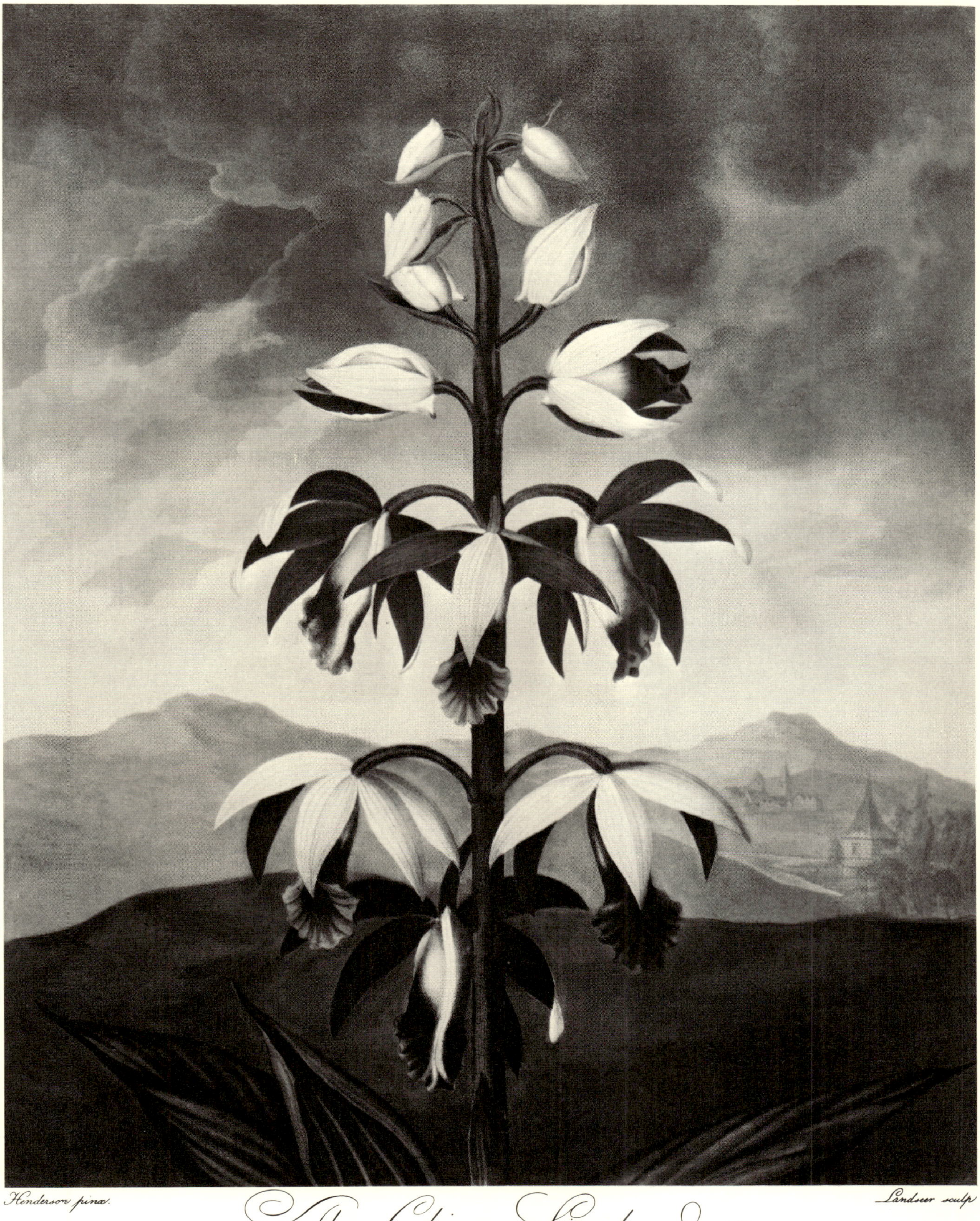

Henderson pinx. *Landseer sculp*

The China Limodoron

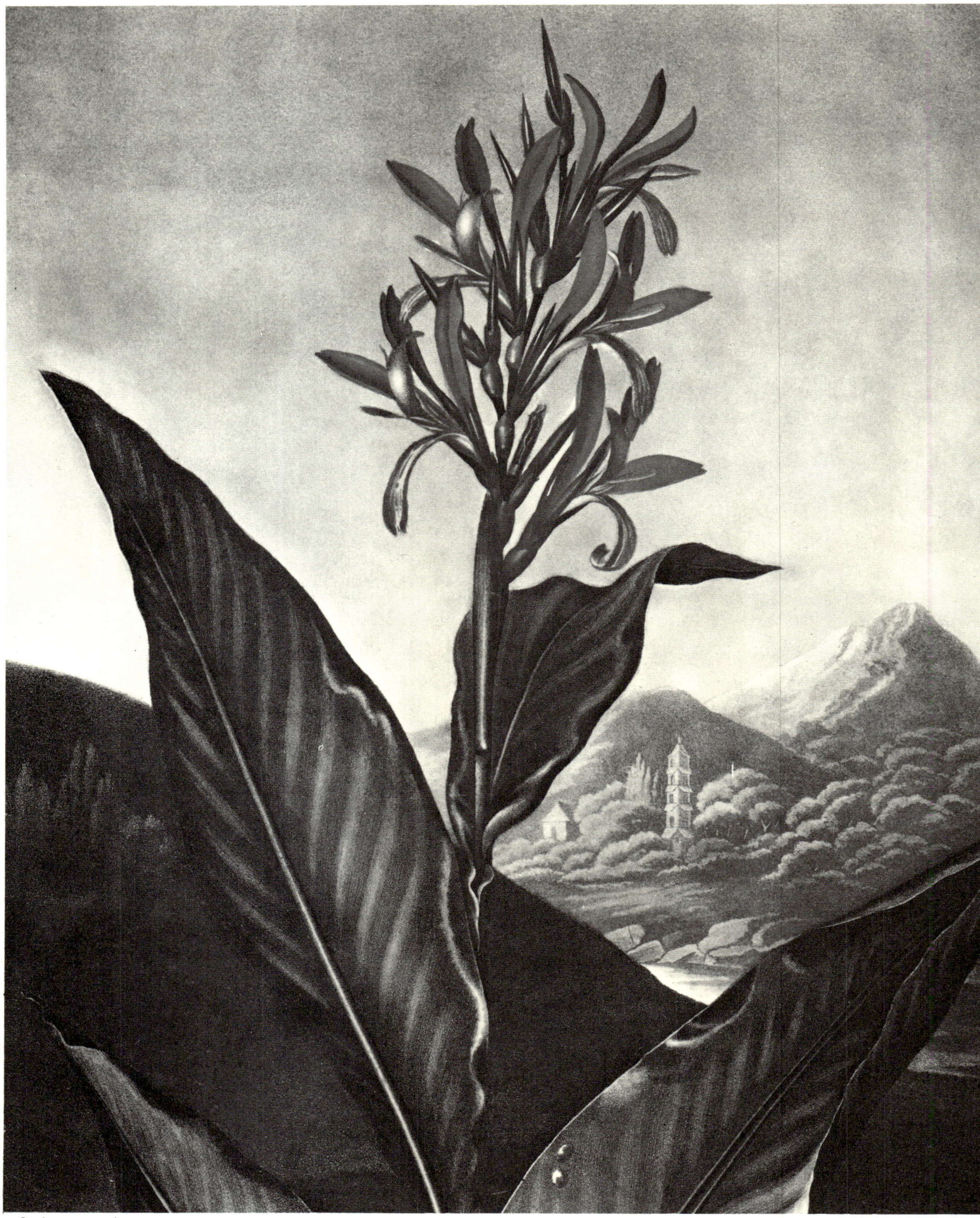

Henderson pinx^t

Caldwell Sculp^t

Indian Reed

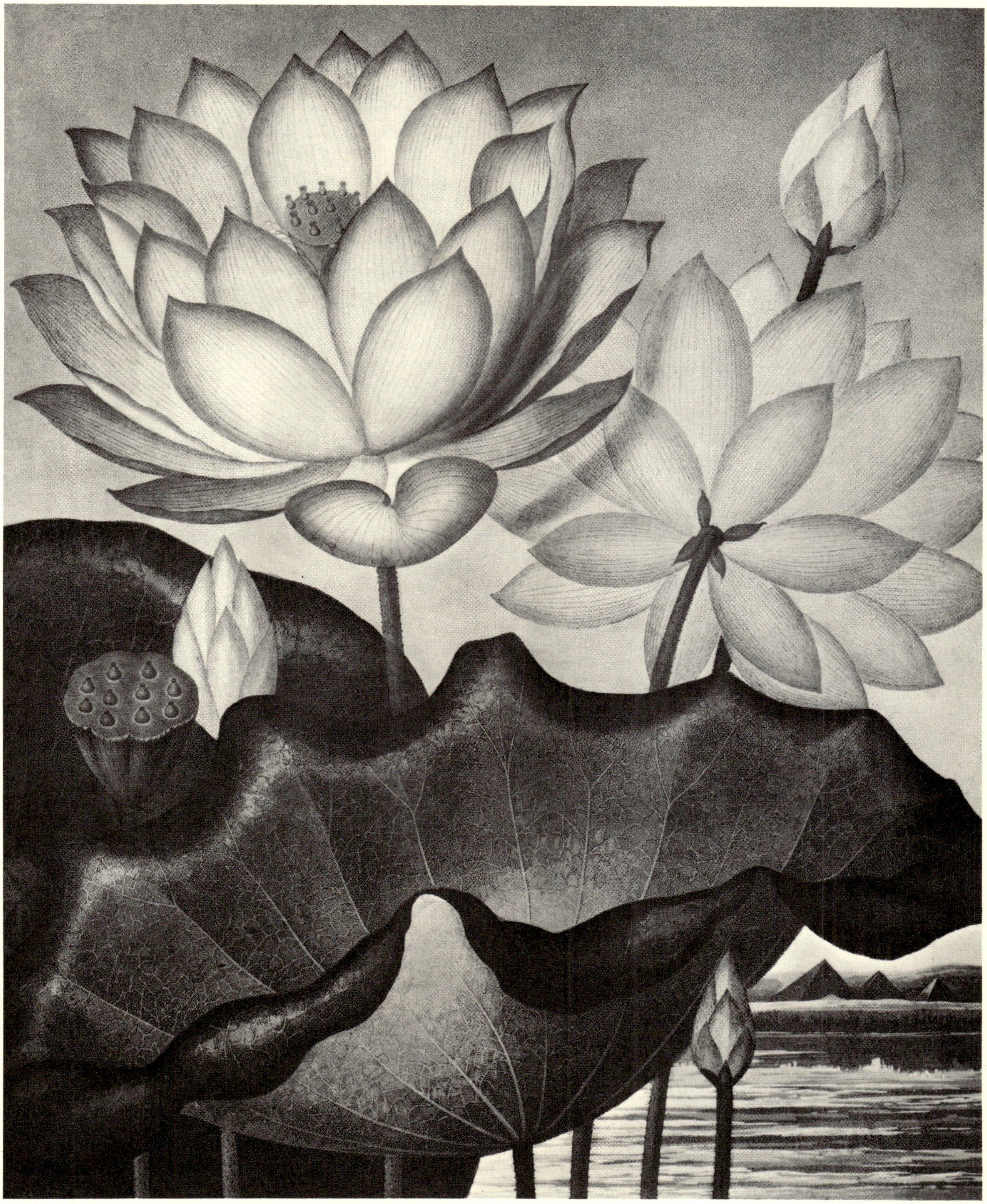

Henderson pinxt — Burke & Lavis sculpt

The Sacred Egyptian Bean

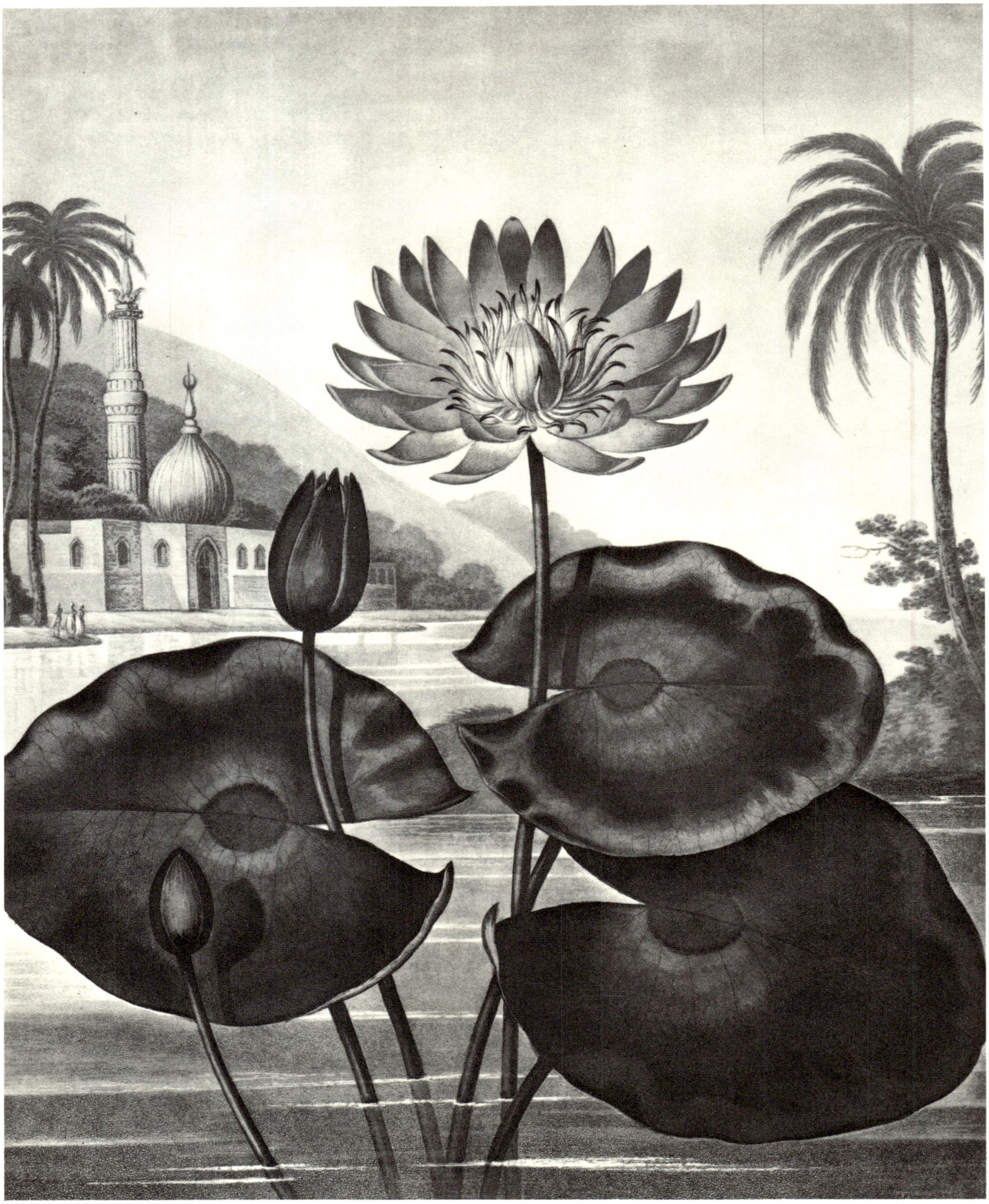

Henderson pinxt — Stadler sculpt

The Blue Egyptian Water-Lily.

BIBLIOGRAPHICAL NOTES

BY HANDASYDE BUCHANAN

The title of Doctor Thornton's whole work, of which *The Temple of Flora* forms the final part, is the *New Illustration of the Sexual System of Linnaeus.* Blunt and Stearn in their *Art of Botanical Illustration* (Collins 1950) says, 'the elucidation of the various editions and issues of this work is a tangle', and this remark applies with equal force to all parts of the book. In my experience, (and I have been lucky enough to see a good many copies of the book and in particular of the *Temple of Flora,* which alone I have got to discuss here) no two copies of this book are alike, and this inevitably means that to some extent the bibliographer's work must be a matter of opinion, and cannot be put forward as other than a point of view. The following notes therefore must please be read with that reservation in the mind of the reader. Much, *very* much, of the analysis of the individual prints in the *Temple of Flora* has been done by Mr Gordon Dunthorpe in his *Flower and Fruit Prints of the 18th and Early 19th Centuries* (Dulau 1938: but unfortunately out of print and very scarce) to whose work I must make the greatest possible acknowledgement, since my catalogue of prints must be based on his; he, perhaps, in his turn must have drawn largely on an article in the *Gardeners' Chronicle* 1894 by Messrs Hemsley and Perkins.

These two works have been my principal bibliographical sources. For the rest I have relied on observation of many copies of the book and even more individual prints. Of copies in private collections which I have been allowed to see, I must specially mention that in the possession of Lieutenant-Colonel P. Bradfer-Lawrence, whose copy has been used for the reproductions in this book, and that of the late Major J. R. Abbey — both of these have very fine early impressions of the prints, and those belonging to Eton College, Dr R. Bevan, and Lord Fairhaven, whose copy is, textually, possibly the completest in existence.

The *Temple of Flora,* of atlas folio size, about 23 inches by 18 inches, was originally issued in parts between 1798 and 1807. In 1807 it was further published in book form with a contents sheet which lists 31 plates – 3 frontispieces and 28 flower plates. In addition to these there was a group of four auriculas as well as the more usual plate with two; a plate of the American Aloe as well as of the Aloe; one of the American Bog-Plants as well as the earlier Pitcher Plant; and a late plate, dated January 1st, 1812, of the Queen Flower which replaced the Queen. Further, there were two quite different plates of the Night Blowing Cereus, though both painted and engraved by the same artists; and two quite different plates of the Superb Lily engraved by different artists. And, finally, two plates from earlier parts of the *New Illustration* frequently make their way into the *Temple of Flora:* the stipple-engraved portrait of Linnaeus by Meyer after Hoffmann and Bartolozzi, which appears in both plain and coloured states; and Linnaeus in Lapland dress, a mezzotint which in its turn was engraved on two different plates, one by Kingsbury, the other by Dunkerton, both after Hoffmann, which also appear both plain and coloured. This gives a total, not counting the uncoloured Linnaeuses, of 40 plates.

Of course no copy that had not been made up later could contain all these prints. The Queen Flower was simply a replacement for the Queen in the lottery issue of the book (described earlier on, by Mr Grigson) which, though issued in 1811, had a title-page dated 1810, and which will be referred to by me as the 1810 issue from now on. And the American Aloe was an unusual replacement of the Aloe. The two plates of the Cereus and the two Superb Lilies never appeared together. Further than this is pure speculation; the Lonsdale copy (now Major Abbey's) contains both Linnaeuses, both Auriculas and the Pitcher and Bog-Plants, and is in contemporary binding, but it is almost certain that this was formed from a collection of the parts. Dr Bevan's copy, which from evidence of watermarks and states of the plates is probably one issued in book form in 1807, contains 31 plates. No 1810 issues ever seem to contain more than the 31 plates plus the Hoffmann and Bartolozzi portrait uncoloured.

This is really all that can usefully be said about the various issues of the book. Estimating the total numbers published is only a little less speculative. We have Thornton's own advertisement of the Lottery as evidence that 199 copies of the 1810 (1811) issue were printed for the Lottery, and in addition 200 portfolios of plates without text. Dunthorne suggests that these greatly outnumbered earlier issues. I cannot agree with this, having seen over a good many years as many of one as of the other, and would suggest that 750–800 copies of the book is a reasonable total figure.

It is now time to consider the much more important question of the individual plates and the prints taken from them. The plates were engraved by various processes – aquatint, mezzotint, stipple and line engraving – and the impressions taken from them were printed in colour and finished by hand. The vast majority of the work on the Thornton plates is done either by aquatint or mezzotint: and most of the plates were altered or added to from time to time, producing a different 'state' in each case, some plates having as many as four different states. The earlier alterations may have been with the idea of improving the plate, but all the later ones, at least, with the sole object of concealing the wear of the soft delicate copper as impression after impression was struck from it.

The actual states of the plates are undoubtedly of considerable technical interest, but do not, in my opinion, materially affect the value of the prints taken from them. They can be detected mainly by two signs: first, the addition of engraved lines to strengthen the work; secondly, by the addition to areas of tone in aquatint, replacing areas of either fine aquatint or mezzotint which have worn.[1]

1 Mezzotint is a process in which areas of the plate are worked with a toothed rocker to give very subtle gradations of tone, showing little texture on the actual print; aquatint is a medium giving coarser effects and may be recognized by the granular texture of its areas of tone.

A number of the plates were never altered, except, of course, that they have worn away. The wear of the plate is an entirely different matter, and that and the excellence of the colouring are the two criteria of both the beauty and the value of the finished print. If the second state results in as attractive a print as the first, as for example in the case of the Blue Passion Flower, which is even more beautiful after the first alterations, then there can be no reason for refusing it. On the other hand, in later states of the Carnations and the Sensitive Plant, where the sky and background have been entirely removed and uninteresting wash substituted, the whole meaning of the print has been taken away and it is worthless from any point of view. In others, such as the American Cowslip, where harsh crude work has been substituted for delicate aquatint, one should again reject it. It is not the addition of a small shadow here, or a few dots there that should influence one's judgement, but signs of wear, which are generally unmistakable and which cannot be covered up with paint, and the colouring. Many different artists must have been employed to do the final hand colouring, and in this as in many other famous colourplate books – notably Boydell's *Thames* – the results in the individual prints vary immensely. It is not too much to say that in this respect no two impressions of any print are quite alike, and really fine copies of the book stand out almost, if not quite, as much for the beauty of the colouring as for the engraving.

The watermark is naturally very helpful, but is not decisive, when one is trying to assess the earliness or lateness of an impression. The watermarks found by myself on prints are Whatman with dates varying from 1794–1810: E & P 1804: and H. Smith 1810. While obviously a late watermark means a late impression, the converse does not always apply. I have found for instance a Renealmia from the 1810 issue with an 1802 watermark, and a certain second state of the Tulip watermarked 1794. No doubt small quantities of the earlier paper came to light later on, and were used as they turned up. But in general the watermark is a very useful guide, with 1794 almost always meaning a first state; there is not usually any need to look for 1810 watermarks, as a glance at the print will show that it has come from sometimes a very worn plate.

The plates vary considerably in the size, the surface enclosed by the plate mark ranging from approximately 14 × 18 inches to 18 × 22 inches. The painter's name in all cases appears in the bottom left-hand corner, the engraver's on the right with the title and date in between. Before proceeding to a short catalogue of the individual plates, I must point out that, without seeing every one of the 800 or so sets of prints, it would be impossible to be certain that all states were included. I am therefore quite deliberately giving the principal details of the changes to be looked for, and am equally arbitrarily refusing to admit to a change of state where a negligible alteration has been made. Dunthorne gives, for instance, a second state of the American Bog-Plant, which I cannot accept, after looking carefully at many copies of the print, since the difference, if any, is trivial. And I have rejected a number of other similar supposed states. But every main alteration will I hope be noted here, and especially such as will enable the reader to judge one print on its own without having others for comparison. I must mention again in this connection that it is the wear of the plate, and the colouring applied to the print, that are the real conditioning factors. It is, incidentally, of interest to notice that the whole 400 impressions of the 1810 lottery edition were struck off without a single alteration in any of the plates, the final alterations, all by this time for the worse, having been made between the edition of 1807 and 1810, during which time of course no prints were issued.

In parenthesis, a quarto edition of the *Temple of Flora* appeared in 1812, illustrated with a Frontispiece of Flora and with 27 plates in aquatint, mezzotint and stipple, some printed in colour, and all finished by hand. The only flower not found in the large edition is the Artichoke Protea drawn by Henderson and engraved in mezzotint by Quilley, showing Table Mountain as a background. Other differences are that, in the Rose plate, the eggs in the nest have hatched out and are now nestlings! And that the Cereus has been entirely redrawn with two blooms. There are no alternative plates and no bibliographical points are raised by the book, which is a pale shadow of its elder brother.

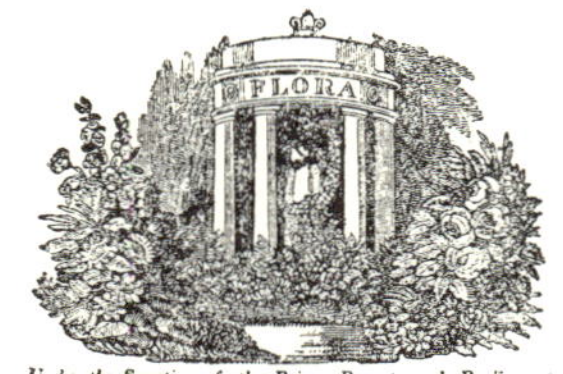

Under the Sanction of the Prince Regent, and Parliament,
CONSISTING ENTIRELY OF BRITISH MANUFACTURE,
A ROYAL
Botanical Lottery,
FOR THE PROMOTION AND ENCOURAGEMENT OF
The Fine Arts, and Science,
Consisting of 20,000 Tickets.

10,000 PRIZES!

ONLY ONE BLANK TO A PRIZE.
To be Drawn early in the Season with the State Lottery.

FOR A WHOLE TICKET. THE PRICE IS NOW
ONLY TWO GUINEAS,
BUT EXPECTED SHORTLY TO BEAR AN HIGH PREMIUM.

This LOTTERY was passed with the Unanimous Consent of PARLIAMENT

BRITONS! join Hand and Heart in promoting the ARTS and SCIENCES of your COUNTRY, by the IMMEDIATE PURCHASE OF A TICKET.
For TWO GUINEAS, a Prize worth above FIVE THOUSAND POUNDS may be obtained

THE LOTTERY PROSPECTUS OF 1812.

CATALOGUE OF INDIVIDUAL PRINTS

These are given in the order in which they appear in the original contents sheet, save that alternative plates are listed next to each other. I have already acknowledged my debt to Dunthorne for his research work on the plates. I have tried always if possible to be a trifle simpler than he is, and often, e.g. on the Rose, have found myself working quite independently. On the other hand, once or twice only, notably in the case of the Blue Passion Flower, I have used his description almost in *extenso*.

I. *Flora dispensing Her Favours on the Earth*

Painter: Cosway. *Engraver:* Woolnoth

Aquatint and stipple

May 1st, 1807. One state only

Flora is sometimes printed in colour, sometimes coloured by hand.

II. *Aesculapius, Flora, Ceres and Cupid honouring the Bust of Linnaeus*

Painters: Russel and Opie. *Engraver:* Caldwell

Stipple. April 1st, 1806
One state only

III. *Cupid Inspiring Plants with Love*

Painter: Reinagle. *Engraver:* Burke

Stipple. June 1st, 1805
One state only

IV. *The Snowdrop*

Painter: Pether. *Engraver:* Ward

Mezzotint, with aquatint added in third state

September 1804

The pond in the right-hand bottom corner is the clue here. In the first state it is entirely open and the water is coloured blue. In the second state bushes growing from the banks almost entirely cover the pond, which looks dark brown. The foreground has also been entirely reworked. Many smaller changes can be seen, notably extra leaves on the creeper above the crocus; the icicles are more prominent; there is more shading on the snowdrops; and the sky has been remodelled. But a glance at the pond will be all that is needed. In the second state the plate has not yet lost all character, though it is much poorer than the first. For the Lottery edition the plate was reworked all over in aquatint, and the crocuses stand out sharply and are crudely coloured; it is hardly worth looking at.

V. *The Persian Cyclamen*

Painter: Pether. *Engraver:* Elmes

Aquatint, stipple and line

1804

There are three distinct states to this plate, which becomes stronger and coarser all the time, yet I have never seen a really ugly print, though each state is worse than the former. This is largely because the colouring has always been skilful. In the first state the top of the castle is indistinct and has no pinnacles on the towers, and this is the first feature to inspect. The hillside is in pure aquatint; the shading behind the cyclamen flowers is lightly cross-hatched, while the tree trunk to the right has only a few lines on it. In the second state the castle is more prominent and five distinct sharp pinnacles have been added, while many extra etched lines are to be seen – notably behind the cyclamen flowers; on the tree trunk; and under the cyclamen leaves on the left, which themselves stand out more sharply. The principal change in the third state is the addition of aquatint to the sky on the left, so that only a streak of light remains above the mountains, while in the earlier states the light reached the top corner. The leaves of the cyclamen are now much coarser, with light and dark patches, and coarse-grained aquatint has been also added to the middle distance. Much additional aquatinting has been applied to other parts of the plate. The most easily noticed differences, however, are the changes in the castle between states one and two, and in the sky between states two and three.

VI. *Hyacinths*

Painter: S. Edwards (Sydenham Edwards, well known for his *New Botanic Garden* and his *Botanical Register*). *Engraver:* Warner

Aquatint, stipple and line

June 1801

This plate must have been reworked early on, as first impressions are very uncommon. It lost a great deal in delicacy afterwards, but the earlier impressions from the second state are still attractive. The plate was finally much worn. The first state is almost all in delicate aquatint with the flowers in stipple. The foreground is hardly lined at all, and is coloured green. The modelling of the river is pleasantly irregular. The clouds are light. A *very* pretty print. In the second state the foreground, now coloured brown, is furrowed with lines. Bushes are added and there is much harsh aquatint. The river has regular, less pleasant lines, while the far bank on the right, which was curving, is now straight. The ship and the house are heavily re-etched. The clouds are also much heavier. The plate is still dignified, but has lost its earlier lightness.

VIII. *Roses*

Painter: Thornton (The Doctor's only painting)
Engraver: Earlom

Mezzotint and line

October 1st, 1805

While there are technically two states of this plate, possibly aesthetically and certainly commercially the most important in the collection, they must be discussed as one. Look at the nest in the foreground: if the individual twigs are clearly distinguishable, down to the smallest tendrils which bind them, the print is from the first state of the plate. Added mezzotint round the edge of the nest later blurred the outlines so as, almost, to form a solid mass round the eggs. With this exception the alterations, *on the plate,* are trivial. Such rework as was done must have been very early on, as few impressions show the nest in its perfection. The colouring, however, varies enormously. Shafts of sunlight over the temple to the top left were painted on to early and late impressions at the whim of the colourist. The earliest impressions of the print have the central rose coloured pink. This then was changed later to a sort of poached egg – orange and red and pink. The plate of course wore steadily as usual, and final impressions are a mere travesty, but, with the single exception of the nest, there is no serious change which can be attributed to the engraver. The bloom on the leaves and the delicate appearances of the various roses are the result of skilful colouring. No plate varies more in this respect, and copies with the 'poached egg' rose, which I attribute, with some diffidence, to the 1807 edition (as opposed to the original parts), still retain a great deal of freshness and charm.

VIII. *A Group of Carnations*

Painter: Henderson. *Engraver:* Caldwell

Aquatint, stipple and line

April 2nd, 1803

Minute alterations which in no way affected the appearance of the prints were made at an early stage of this plate, which so far is one of the finest of Thornton's productions. In its final state, however, which appears only in the 1810 edition, the background has been entirely removed and a bluish-white wash substituted, while the temple on the right has been re-engraved. The whole plate has worn so that it is scarcely recognizable, and the colouring has been put on carelessly. Impressions of the Carnations vary therefore more widely than any others in the book, but equally there is probably less difficulty than in the case of any others in distinguishing between good and bad. A small apparently discoloured patch where the ink has run on the right-hand blue carnation is present in all early impressions.

IX. *A Group of Auriculas* (four)

Painter: Henderson. *Engravers:* Lewis and Hopgood

Aquatint, stipple and line

May 1st, 1803

First state: delicate aquatint sky. Three eagles at top centre of plate. The mountains stand out well. In later impressions the eagles become much fainter.

Second state: the eagles have been removed; the sky has added coarser aquatint, this has also been added to the mountains which are less distinct, and darker.

This plate is not often found. Although first impressions are more attractive, the difference first to last is not so pronounced as in most other Thornton prints.

IXA. *A Group of Auriculas* (two)

Painter: Reinagle. *Engraver:* Sutherland

Aquatint

May 1st, 1807

Two distinct states, but as in the case of the four auriculas the character of the plate is not changed entirely. Final impressions are, of course, worn, and much painted over. All the plate in the first state is in delicate aquatint and the fir trees on the mountain are hardly visible. In the second state, coarse aquatinting behind the first makes them stand out more sharply, as does the outline of the mountain itself, while the foliage and flowers of the auriculas have been reworked.

X. *Tulips*

Painter: Reinagle. *Engraver:* Earlom

Mezzotint

May 1st, 1798

This was the first print to be published by Thornton and two trial proofs of it exist, in the Josi collection, which was presented by Dr Thornton as the finest and largest possible collection of his prints. One of these proofs was lettered below in upright instead of the usual sloping capitals, and traces of this can be found under the lettering in the ordinary issues. These traces are in no sense the hallmark of early impressions but are visible equally in early and late.

In the first state of this plate the windmill in the middle is scarcely visible, while the fields in the foreground are undefined. The church and sky to the right over it are soft and charming.

In the second state the fields are reworked and show up clearly; bushes are added. The windmill stands out and the church has been brought out plainly but now rather meanly. The sky above is reworked and less attractive. As the plate wore the fields again became less clear and more like the first state. So look first at the church and the windmill.

Unlike many other of the prints, late and early impressions seem to be well coloured and I have never seen one that was not attractive.

XI. *The Queen*

Painter: Henderson. *Engraver:* Cooper

Stipple and line

February 1st, 1804. One state only

XIA. *The Queen Flower*

Painter: Reinagle. No engraver's name

Aquatint, stipple and line

January 1st, 1812. One state only

A late substitute for the Queen, and in every way worse.

XII. *The Aloe*

Painter: Reinagle. *Engraver:* Medland

Aquatint and line

May 1st, 1798

There are two states of this plate, but while the superficial differences are small, their effect is considerable, and the first state is much to be preferred. This is partly due to the colouring, a delicate, pinky tinge in the sky having disappeared in the second issues; and partly also due to comparatively early wear, which was probably responsible for the introduction of the alternative plate of the American Aloe.

The principal changes in the second state are a certain amount of rework on the foliage of the Aloe, and on the sky, while the horizon line is now hard and ugly.

XIIA. *The American Aloe*

Painter: Reinagle. *Engraver:* Medland
Aquatint and line
May 1st, 1807
One state only
A substitute for the Aloe, only found in later copies of the book.

XIII. *The Nodding Renealmia*

Painter: Henderson. *Engraver:* Caldwell
Aquatint, stipple and line
November 1st, 1801
This is a plate which in its later stages wore very badly indeed, and final impressions have lost all the charm of the early ones which are magnificent.

There are two states of the plate, the principal changes in the second being the addition of much coarse aquatint to the foreground, while the outline of the mountains has been strengthened. Not, however, a heavily reworked plate.

XIV. *The Night Blowing Cereus*

Painters: Flower by Reinagle. Moonlight by Pether. *Engraver:* Dunkerton
Mezzotint
May 20th, 1800
There are two quite different plates of this painting, both engraved by Pether and both used at the same time, which I shall call A and B. A was, I think, the more used, and late impressions are rather worn, but neither plate was seriously affected by the later work done on it, nor is A or B noticeably the better plate. The principal differences between them are as follows:

A. The flourish of the T of The in the title cuts through the ht of Moonlight in the left-hand corner. The flourish of the N of Night has a loop. The minute hand on the clock covers the dot between XII and I, while the clock has no band in which the hour figures are placed. There is a dark spot in the centre of the Cereus. The plate is slightly different in size of engraved surface from B and the length of the title is 9¾ inches.

B. The flourish of the T of The does not touch the word Moonlight. The flourish of the N of Night has no loop. The dot between XII and I is visible and the clock has a band for the hour figures. The centre of the Cereus has no dark spot. The length of the title is 8½ inches.

In the second state of A there is added mezzotint on the petals of the bloom and the spines on the stem of the Cereus. The clock face is darker and the ripples in the water are wavy and not straight as before.

In the second state of B, a number of etched lines are added to the plate strengthening the foliage of the trees behind the flower, the outer petals of the blossoms, the bank above the water and the trees on the left. Both plates are very fine and really bad impressions are few, no doubt owing to there being two plates.

XV. *The Oblique-Leaved Begonia*

Painter: Reinagle. *Engraver:* Caldwell
Aquatint, stipple and line
July 1st, 1800
A very fine plate at first, with the begonias actually all printed in colour in the early impressions, but it wore out entirely in the end and became one of the most pale and lifeless of them all. Minor variations in state only. The mountains which start in pure aquatint are later strengthened with oblique lines, while the horizon is sharper. Later still, the large white cloud at the top right became reduced in size from 1–1½ inches to less than ½ inch, though with final wear it becomes a shapeless 1½ inches again.

This print, together with the four auriculas and the Indian Reed, is often found with peculiar oxidation marks particularly at the top and right side.

XVI. *Large Flowering Sensitive Plant*

Painter: Reinagle. *Engraver:* Stadler
Aquatint, stipple and line
December 1st, 1799
I have found only two states of this plate. Dunthorne gives a third between my two, and I include his principal differences. This is the only state in the whole catalogue which is not described from personal observation. Impressions from the first state of the plate are very fine – among the most delicate of all the prints. The sky is worked in fine aquatint, which in the *final* state, as in the case of the Carnations, has been entirely removed, thus altering once again the whole balance of the picture. In this state, two small circles have been added at the ends of the stamens.

In Dunthorne's second state, the shoot at the top end of the stem of the sensitive plant has its two tips joined in a point (this should be easily noticeable), while the buds and leaflets have been outlined more clearly. The sky has been reaquatinted with streaks, and the background too has added aquatint.

XVII. *The Blue Passion Flower*

Painter: Reinagle. *Engravers:* Caldwell
Aquatint, stipple and line
January 1800
Three very definite states, the second of which is, I think, actually better than the first. The final state shows heavy rework almost everywhere and the plate has been ruined. As the alterations are all minor, between one and two, it is perhaps worth giving them in full.

First state: the main stem and the stems of the bud and the leaf in the upper centre have a pronounced shadow with hard outlines. The dark patch of shading to the left of the base of the columns has a hard upper edge. The part of the leaf to the right of the fruit is in aquatint only and so is the leaf projecting to the left beyond the column. The dark shading to the right of the column above the leaf has a jagged edge and the cloud above has an indistinct edge.

Second state: the shadows of the stems in the upper centre are smaller and the edges are softer. The dark patch of shading is softened. The aquatint on the leaves mentioned above is strengthened with lines, as is the shading to the right of the column. The shadows of the leaves in front of the lower part of the column have been strengthened with cross-hatched lines. There are added horizontal lines on the base of the column near the fruit.

Third state: coarse aquatint on the leaves and buds; the indistinct cloud has now a white edge, and there is heavy line rework on many of the leaves. The flowers are no longer all printed in colours as before.

XVIII. *The Winged Passion Flower*

Painter: Henderson. *Engraver:* Warner
Aquatint, stipple and line
June 1st, 1802
Two states

In the first, the shadow of the vine extends to the top of the plate. The top of the column is solid, with no fluting, while the shadow under the top leaf is smooth.

In the second state the shadow stops well before the top of the plate. The fluting has been put in at the top of the column. The shadow under the top leaf is jagged. The foliage on both sides of the column has been reworked. This second state was very good at first but later wore very much, so that it is difficult not to think that there has been a third state; but all changes are the result of wear.

XIX. *The Quadrangular Passion Flower*

Painter: Henderson. *Engraver:* Hopwood
Aquatint and stipple
June 1st, 1802
Three states

In the first, the fluting at the top left of the column is scarcely visible, while there are clouds on either side of the column. In the second, the indistinct fluting has been made clear, while a shadow has been added in aquatint above a tendril just below the capital of the column. The clouds have almost gone. In the third, much coarse stipple has been added over the aquatint in many places and especially on the leaves, and the base of the column is more distinct than in the earlier states. The temple and foliage to the right have been reworked. The first state of this plate is much the best, though it had begun to deteriorate early on, so that even the second state shows signs of wear.

XX. *The White Lily*

Painter: Henderson. *Engraver:* Stadler
Aquatint
August 1st, 1800

Four distinct states, but only the fourth presents an aesthetically inferior effect. The first three have interesting but quite unimportant differences, nor does the plate appear to have worn much. The feature to look at is the temple on the right. In the first state, the shadow on the columns runs down obliquely from the left, but that on the second column is 1/4 inch higher than the other two; further, there is a *slightly* visible clump of trees to the left of the temple, in front of the distant hill.

In the second state, the shadow on the second pillar has been entirely removed; so has the clump of trees. In the third state, the shadow on the column has come back, but runs down symmetrically with those on either side of it, while the clump of trees has come back and is clearer than in the first state. These alterations, though small, are visible at a glance; to list other equally trivial points would only confuse.

In the fourth state, however, the plate is much worn and no longer good. The shading on the blossoms has worn away and coarser work has been substituted, and the background has been reworked showing five distinct rock formations not previously visible.

XXI. *The Superb Lily*

Painter: Reinagle

There are two entirely different plates of this painting, one engraved by Ward, the other by Earlom. Both were engraved in mezzotint, with added aquatint to the later states of the Earlom plate. The date is June 1st, 1799, and both plates seem to have appeared simultaneously, as impressions of each with 1794 watermark exist. There is little difference aesthetically between the two.

The Ward plate had one state only, and probably was used for the parts alone. I have never seen a print that was not faultless. The Earlom plate had three states, the third of which wore very badly indeed, so as to present almost the appearance of a fourth state.

The first state is in pure mezzotint and rivals the Ward plate. In the second state a very little aquatint has been added to the sky and the lily and the trees in front of the mountainside. In the third state, aquatint has been added *all over the plate.* The lily is now rigid and stiff as are the trees in the background, and the foreground which before was indistinct has been broken up. The whole plate is harder, and indeed produces quite a different picture. And yet, till the plate wore badly, which it did not do at once, it remained very attractive. Some critics, indeed, prefer the third to the first states of this plate, though I cannot agree with them.

XXII. *The Dragon Arum*

Painter: Henderson. *Engraver:* Ward
Mezzotint, aquatint added in final state
December 1st, 1801

Three states: there is little resemblance between the start and finish of this plate.

First state: the sky at the right has a definite mountain peak between the storm clouds which are carried across the top of the plate.

Second state: the top half of the plate has added mezzotint, which gives a curious effect like superimposed wire netting. The dark cloud at the top reaches to about the middle of the plate.

Third state: added aquatint in foreground and middle distance. The sky on the right is pierced with flashes of lightning, and the mountain has become an erupting volcano with small clouds round its base. The sky at the top is brighter and coloured yellow.

All three states are found in reasonable condition, and the third is more exciting though really less good than the first; but the plate as usual wore badly at the end, and the colouring was brightened, sometimes grotesquely – a pale blue line painted up the black centre of the Arum producing an effect of the Old Etonian colours!

XXIII. *The Maggot-Bearing Stapelia*

Painter: Henderson. *Engraver:* Stadler

Aquatint, stipple and line

July 1st, 1801

Two not very different states

In the first, the mountain behind is wooded and has clouds above it. Both trees and cloud have disappeared in the second. There are other signs of rework, but they do not greatly affect this rather unimportant plate.

XXIV. *Pitcher Plant*

Painter: Reinagle. *Engraver:* Cooper

Aquatint, stipple and line

November 1st, 1803. One state only

A second crane to the right of the one in the foreground has been partially scraped off the plate, but the outline remains clearly.

XXIVA. *American Bog-Plants*

Painter: Reinagle. *Engraver:* Sutherland

Aquatint

July 1st, 1806. One state only

An alternative plate to the Pitcher Plant.

XXV. *The Pontic Rhododendron*

Painter: Henderson. *Engraver:* Caldwell

Aquatint, stipple and line

December 1st, 1802. One state only

The colouring of this print varies considerably, irrespective of date of the impression, but, as it wore considerably, earlier issues are much prettier.

XXVI. *The American Cowslip*

Painter: Henderson. *Engraver:* Warner

Aquatint, stipple and line

May 1st, 1801

Three states, of which the second is only slightly less attractive than the first.

In the first state, there is only one patch of blue sky over the nearer of the two ships, which together present an appearance of perspective. The foreground changes gradually from dark at the bottom to lighter under the rocks, which are more or less unbroken.

In the second state, coarse tufts of grass stand out sharply in front of the rocks at the back of the foreground, while bushes growing out of the rocks are enlarged. This is not, however, a great change, though the plate is already coarser.

In the third state, the plate has been reworked almost all over.

The ships look tawdry, with lost perspective, and there are five blue patches of sky over the first ship. The foreground now shows alternate light and dark patches, and the crevices in the rocks show up much more.

In the earlier states, the plate has a depth and mystery entirely lacking in the third.

XXVII. *The Narrow-leaved Kalmia*

Painter: Reinagle. *Engraver:* Caldwell

Aquatint and stipple

June 1st, 1804. One state only

XXVIII. *The China Limodoron*

Painter: Henderson. *Engraver:* Landseer

Aquatint, stipple and line

February 1st, 1802. One state only

In later, worn, impressions the village on the right has almost disappeared.

XXIXX. *The Indian Red*

Painter: Henderson. *Engraver:* Caldwell

Aquatint and line

1804

First state: the lower part of the sky is in aquatint which reaches to the top of the mountains. Painted by hand on the two flowers are five additional petals and a long pointed bud at the top of the spike of the bloom. This painting has, of course, nothing to do with the state of the plate, but does, in fact, appear only on the earlier issue.

Second state: the aquatint in the lower part of the sky has been removed, leaving a clear washed band of sky above the mountains. The hand colouring mentioned above is not there, hence the flowers are smaller. The leaves of the reed have a peculiar withered appearance at the edges. The whole effect of the plate, background and foreground, is now ruined.

XXX. *The Sacred Egyptian Bean*

Painter: Henderson. *Engravers:* Burke and Lewis

Aquatint and stipple

December 1st, 1804. One state only

XXXI. *The Blue Egyptian Water Lily*

Painter: Henderson. *Engraver:* Stadler

Aquatint

September 11th, 1804. One state only